Introductory Physics

Introductory Physics

J. W. Hill

Head of Science Faculty,
Oak Park Comprehensive School,
Havant, Hampshire

Macmillan
Education

First published 1980
Reprinted 1981

Published by
Macmillan Education Limited
Houndmills Basingstoke Hampshire RG21 2XS
and London
Associated companies in Delhi Dublin
Hong Kong Johannesburg Lagos Melbourne
New York Singapore and Tokyo

Printed in Hong Kong

British Library Cataloguing in Publication Data

Hill, James William
Introductory physics.
T. Physics
I. Title
530 QC21.2

ISBN 0-333-24442-7

Contents

Preface

This book is a foundation course in physics bridging the gap between combined or integrated science courses in the early years of secondary school, and the examination-centred work of subsequent years. Its aim is to provide a basic grounding in physics essential both for pupils who will continue to single-subject physics examinations, and for those who will take their study of physics no further.

No practical work is included, and the book should be seen, therefore, as complementing a scheme of laboratory-based activities. Strong emphasis is placed on straightforward presentation of concepts with the minimum of quantitative work. Calculations, where used, are set out in a detailed step by step fashion. Self-tests and question sections at the end of each chapter reinforce and amplify the physics covered, developing the vocabulary, knowledge and confidence, essential to any successful foundation course.

Acknowledgements

The author and publishers wish to acknowledge the following photograph sources:

British Museum (Natural History) p. 14 top right;
Reproduced by permission of the Director, Institute of Geological Sciences; NERC Copyright reserved pp. 14 top left, 14 bottom right;
Graham Photography p. 56;
Philip Harris Ltd p. 16;
C James Webb pp. 18, 24, 35, 82, 103.

The publishers have made every effort to trace copyright holders, but if they have inadvertently overlooked any they will be pleased to make the necessary arrangements at the first opportunity.

Introduction

Physics is the branch of science which deals mainly with the study of matter and energy.

By matter we mean materials which can be examined by measuring, by weighing and by other experimental testing. We can weigh and measure solids, liquids and gases but such things as thoughts, ideas and feelings cannot be treated in the same way. These do not belong to the 'material' world.

Energy is vital to life: it enables us to move, to 'work' and to bring about all sorts of changes in the position and the condition of matter. Heat is a form of energy and so is electricity.

Physics enables us to find out about the 'natural laws' which operate in the universe. These are not the same as man-made laws drawn up, for example, by parliaments to control the actions of people and with penalties for those who break them. A natural law might be described as a fixed pattern of behaviour: the north pole of a magnet will always attract the south pole of another magnet and repel the north pole. Natural laws do not stop us from achieving our ambitions, such as to put a man on the moon, but we have to take them into consideration. Whatever we do has to be done within the framework of the natural laws.

Knowledge of natural laws has been gained largely by experimental work and by careful thinking. An experiment is really 'trying something out'. A good experiment is one which is set up with regard for safety and arranged so that as many things as possible which are likely to affect the result are kept under control. A good experimenter keeps an accurate record of 'what was done' and 'what happened'. A good record is one which uses few words but uses these carefully. Diagrams of apparatus save descriptions: numbers and calculations are really a precise way of making descriptions.

From experiments, a 'theory' of *why* things happen may be worked out. Further experiments are carried out to test the theory and to see if it is reliable. A theory is a reliable one if it is able to *predict* what the outcome of an experiment will be.

This book is intended to be an introduction to the world of physics. Physicists are working all the time to find out more about matter and energy. Nowadays, their interests often seem to be very different from the basic ideas of physics, but even the most complicated work that they are doing has been made possible by the work of physicists of the past. The knowledge contained in this book was gained by scientists of the past, working, in the same way as scientists do today, with enquiring minds and the ability to set up experiments and make careful observations of the results and to test theories patiently and honestly.

1 Materials

Here is a list of some common MATERIALS:

water air wood iron glass milk aluminium
rubber granite oxygen petrol ice copper
nylon hydrogen steam paper vinegar

Most of these can be touched or handled. Samples could be weighed to find out if they are 'light' or 'heavy'. This is difficult in the case of gases which are often invisible, like air, oxygen and hydrogen, but these may be sniffed (with caution) to see if they have a smell, tested with a flame to see if they burn, and squeezed to find out if they can be compressed. All materials can be measured in some way to find out how much space they take up, how heavy they are and how they behave when heated or when connected to electricity.

One scientific method is to CLASSIFY them, to put them into groups if they are in some way similar to each other. A simple classification of the materials in the list could be:

SOLIDS	LIQUIDS	GASES
wood	water	air
iron	milk	hydrogen
glass	petrol	oxygen
aluminium	vinegar	steam
rubber		
granite		
ice		
copper		
paper		
nylon		

Solids A solid is something which has a shape and will keep that shape unless something is done to it, for example if it is cut, sawn, crushed or heated strongly. Even materials which are normally found as powders or crystals, like flour or sand, are solids.

Liquids Liquids do not have a shape but take up the shape of the container they are kept in. Out of the container, liquids flow and cover as wide an area as possible. (See Figure 1.)

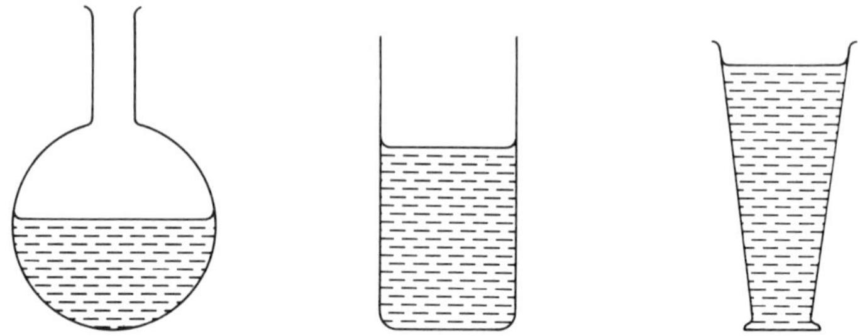

Figure 1 A liquid takes up the shape of the container in which it is placed

Gases

Gases are very light compared with solids and liquids and are often invisible. Like liquids they have to be kept in containers but the containers must be closed otherwise the gas will escape into the air. When in a container, a gas will spread through all the space that is available and will not stay in the bottom like tea in a teacup. Gases are easily 'compressed' by being squeezed to take up less room. (See Figure 2.)

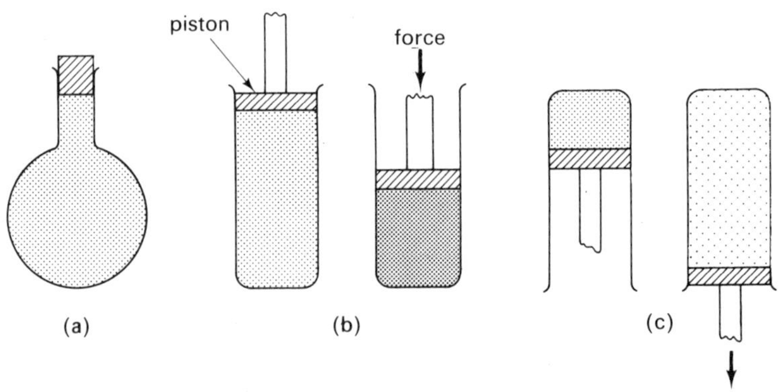

Figure 2 (a) Gas in a container spreads out to fill the whole space evenly. (b) When compressed, the gas will be squeezed into less room. (c) Given more room, the gas will spread out to fill the extra space evenly

Ice is a solid. *Water* is a liquid. *Steam* is a gas.

By steam is meant the invisible material over the surface of boiling water, not the misty cloud which can be seen further away from the surface. Although this cloud is often called steam, it is made up of tiny drops of water formed when true steam, or water vapour, cools or CONDENSES. (See Figure 3.)

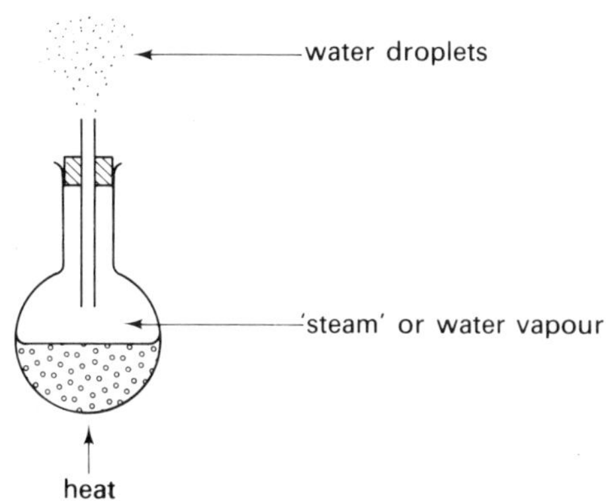

Figure 3

Water (a liquid) is made into ice (a solid) by freezing and into steam (a gas) by boiling. All three are the same MATERIAL in different STATES. Lead may be made into a liquid by heating and when cooled it returns to the solid state.

Solid, liquid and gas are the three STATES of matter and many materials can be made to exist in these three states.

Metals

In the list of materials at the beginning of this chapter there are some METALS: iron, aluminium and copper. If someone were given samples of different materials and asked to pick out the metals, he might say that this could be done with a magnet, thinking that the magnet will attract the metals. This is not so; a magnet will attract only *certain* metals. These are iron, steel, nickel and cobalt: a magnet will not attract copper, aluminum and lead nor any other metal. How is it possible to decide whether a material is a metal or not? Metals have many common features but there are some exceptions to each:

(i) Metals are nearly all 'heavier' than water. (Sodium is an exception since it floats and reacts chemically.)

(ii) Metals are normally solid but can be melted when heated. (Mercury is a liquid at normal temperatures but becomes a solid when cooled to a temperature far below the freezing point of water.)

(iii) Metals allow an electric current to flow through them: they are electrical conductors.

(iv) Metals allow heat to travel through them fairly easily: they are good heat conductors.

(v) Metals can be made to shine and reflect light. A metal may have a dull surface but if scratched with a file it soon shows a 'metallic lustre'.

Materials which are not metals are simply called non-metals.

Figure 4

Crystals

Many solid materials exist as crystals. If samples of common salt and sugar are looked at under a microscope or a lens, it can be seen that the small particles are all very similar in shape and resemble tiny cubes. A crystal shape is always made up with flat sides and straight edges. It is not often possible to pick out a perfect crystal shape from crystalline substances as the edges tend to get broken and crystals get stuck together. Perfect crystal shapes can be 'grown' experimentally. Figure 4 shows some typical crystal shapes.

Elastic materials

Rubber is a solid but it is also a material which can be bent or stretched and will go back again to its original shape provided that it has not been bent or stretched too much. An elastic band is an obvious example of something like this. 'Elastic' is not really a material but it is a word which describes materials which have the property of going back to their original shape. Many metals are springy in this way. Gases have an elastic property. If a finger is placed over the end of a bicycle pump and the air in the pump is then compressed, when the force pressing the pump handle is released the air will immediately 'spring back' to its original size.

Self-test

SOLID LIQUID GAS CRYSTALS ELASTIC COMPRESSED STRETCHED CONDENSES STATES CONDUCTOR HEAT LUSTRE METAL NON-METAL

Using the words in the list above, copy out and complete the sentences A to G:

A Materials may exist in three ______ . These are ______ , ______ and ______ .

B A material through which an electric current can be passed is called an electrical ______ .

C All metals will allow electricity and ______ to pass through them quite easily.

D When steam is cooled it ______ into water.

E ______ have flat sides and straight edges.

F A material which can be ______ or ______ but which will spring back into shape when the force is removed is called an ______ material.

G All metals when polished reflect light and have a metallic ______ .

Questions

1 Classify each material in the following list as a metal, non-metal or gas:

rubber magnesium helium nylon nickel chlorine
carbon silver air sugar sulphur chromium
polythene zinc chalk aluminium salt

2 From the list of materials in question 1, pick out those which
(i) have elastic properties
(ii) form crystals
(iii) conduct heat and electricity well.

3 What are the main differences which lead you to decide if a material is a solid, a liquid or a gas?

2 Volume, mass and density

Figure 5 shows a brick and a piece of wood. The brick is 'bigger' than the block of wood because it takes up more space. How much space an object or a certain amount of material takes up is called its VOLUME. The volumes of shapes like cubes may be worked out if the length, width and height are measured with a ruler and the figures are then multiplied together to give the volume in cubic units.

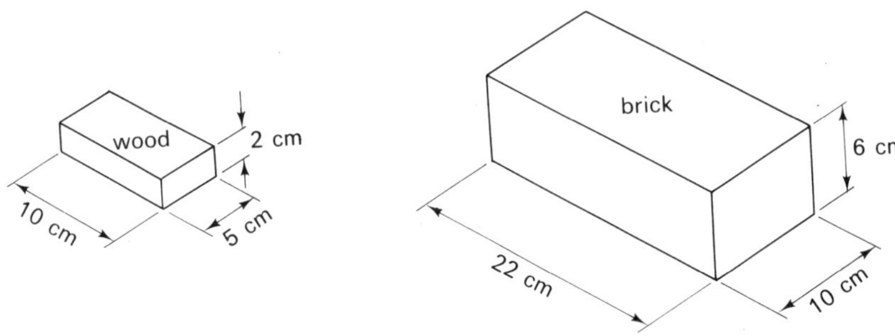

Figure 5

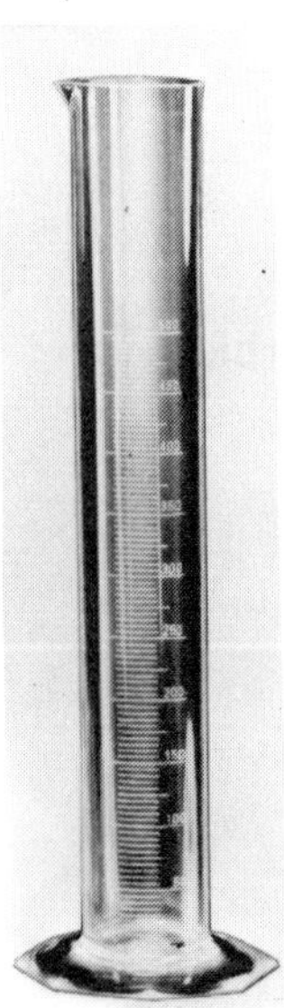

Figure 6 Measuring cylinder

For example the brick measures 22 cm by 10 cm by 6 cm. Its volume is:

$$22 \text{ cm} \times 10 \text{ cm} \times 6 \text{ cm} = 1320 \text{ cm}^3$$

The expression cm^3 is a short way of writing centimetre cube (a cube which measures 1 cm × 1 cm × 1 cm). Where very big volumes are measured, the unit is m^3, a cube measuring 1 metre along its sides.

One way of finding the volume of a quantity of liquid requires a piece of apparatus called a measuring cylinder. This may be marked off (CALIBRATED) in cm^3. Some measuring cylinders are marked ml. 1ml is a millilitre, a thousandth part of a litre. Although it is very slightly different it may be taken to be the same volume as 1 cm^3.

When liquids are poured into a measuring cylinder, or any vessel, the surface of the liquid is attracted to the glass and the surface forms a slightly curved shape called a MENISCUS. The measurement should be taken by reading from the bottom of the meniscus. Mercury, which does not 'stick' to glass, forms a differently shaped meniscus. (See Figures 6 and 7.) The height of mercury in a tube should be read as shown in Figure 7(b).

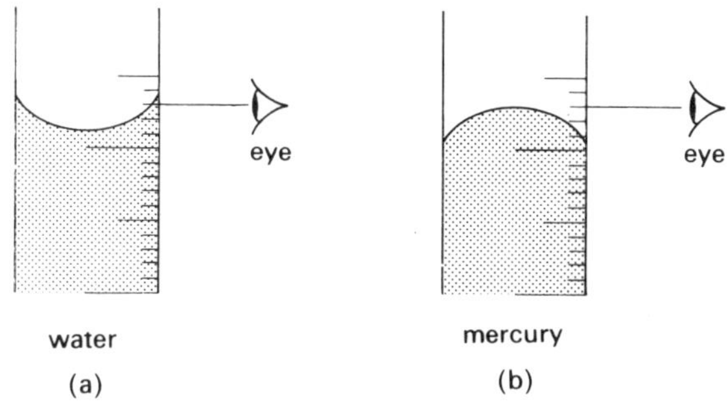

Figure 7 The meniscus

To find the volume of a solid object, such as a pebble, which has an irregular shape, a piece of apparatus called a displacement vessel is needed (Figure 8). This is filled with water until the water runs out of the spout. It is left until no more water runs out. An empty beaker is then placed beneath the spout and the pebble is lowered into the displacement vessel on a thin thread. The water displaced by the pebble is collected in the beaker and its volume is then found by pouring it into a measuring cylinder.

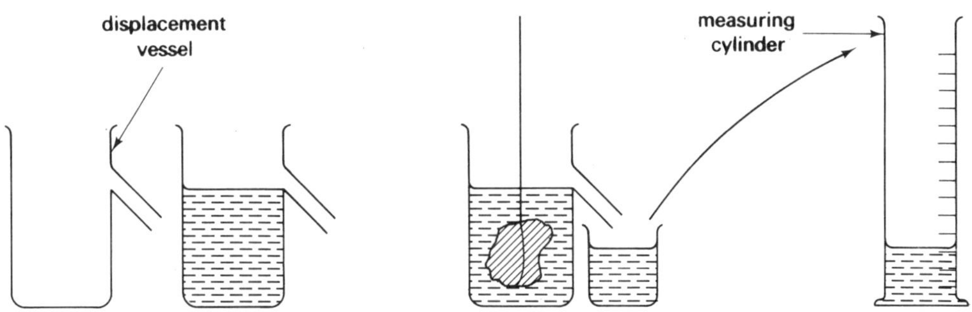

Figure 8

To find the volume of objects which float in water the following method may be used:

(i) Attach something heavy to the light object so that it will sink.
(ii) Find the volume of both the objects when tied together.
(iii) Find the volume of the heavy object on its own.
(iv) Subtract the volume of the heavy object from the volume of both the objects together.

The volume of a certain quantity of gas is not simple to measure. The volume of a gas depends upon the pressure and on the temperature at the time of measurement. A certain volume of gas may be collected as shown in the diagram (Figure 9).

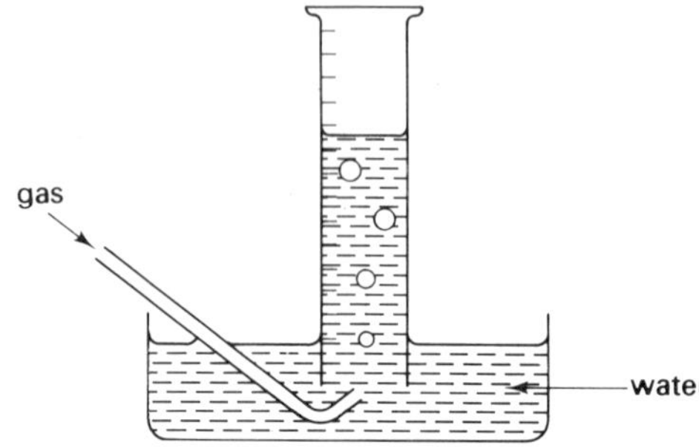

Figure 9

The tall cylinder is filled with water and a ground glass plate is placed over its mouth. The cylinder of water is turned upside down keeping the plate in position and put into the trough of water. The plate is removed when the mouth of the cylinder is under the surface and the water stays in the cylinder (the cylinder is hand-held in position). When gas is bubbled in through the tube, the water is displaced. The volume of gas collected will get larger if the temperature increases and will become less if the pressure of the atmosphere increases. The gas in the cylinder will also contain some water vapour.

Mass

Mass is a word used frequently in physics where we use the word 'weight' in everyday life. When we use the word weight we are explaining 'how heavy' anything is. In physics, WEIGHT has a special meaning: it means the measurement of the force of gravity acting on

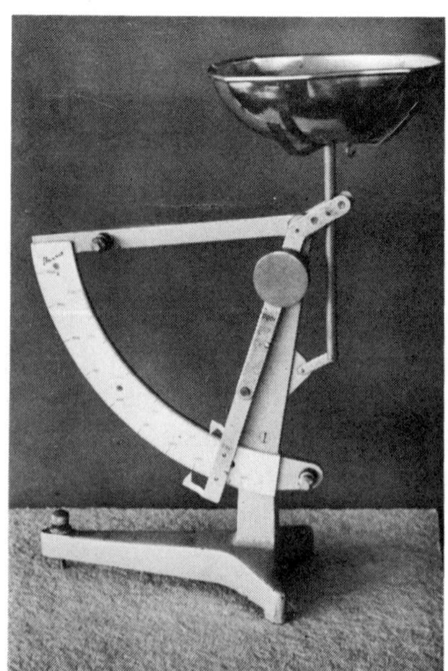

Figure 10

anything and this may be different according to the place where the weight is measured. A man on the moon weighs about one-sixth of what he does on Earth because the pull of gravity on the moon is less than that on Earth. But the man has the same MASS: he contains the same amount of matter. More will be said about weight and its measurement in chapter 3.

Mass is usually measured in the laboratory by using a 'quick-weighing' balance or a top-pan balance (see Figure 10).

The units of mass are the kilogram (kg) and the gram (g); 1000 g = 1 kg.

Density

A block of aluminium has a mass of 250 g and a block of iron has a mass of 150 g. It would be wrong to say that aluminium is a heavier *metal* than iron. The block of aluminium with a mass of 250 g would be bigger than the block of iron with a mass of 150 g. In other words, the block of aluminium has a greater volume. The 'heaviness' of the metals can only be compared if *equal volumes* are tested. In this case it is fairly obvious that aluminium is 'lighter' than iron: aluminium is used for making aeroplanes but it is more accurate to say that aluminium is *less dense* than iron. In the same way, hydrogen gas which is used for filling ballons, is *less dense* than air.

In elementary work, density is measured by finding the mass in grams of 1 cm^3 of a material but density is frequently given as the mass (kg) of 1 m^3 in reference books.

To find the density of iron and aluminium it would be possible to make a centimetre cube of each metal and then find the mass of each, but this would take a lot of time and it is not necessary. If we find the mass of the piece of aluminium and then its volume, the density can be calculated by dividing the mass by the volume. The mass may be found by using the balance and the volume found by using a displacement vessel (see chapter 1).

$$\text{density} = \frac{\text{mass}}{\text{volume}} \qquad \text{or} \qquad d = \frac{m}{v}$$

(The mass must *always* be divided *by* the *volume*: there is a temptation to turn the sum round the other way if the number showing the mass happens to be less than the number showing the volume. In such a case the density is less than 1 g/cm^3.)

Two examples of density calculations might appear as follows:

(a) *Density of aluminium*
Volume of aluminium = 40 cm^3
Mass of aluminium = 108 g

$$\text{Density of aluminium} = \frac{108}{40} = 2.7 \text{ g/cm}^3$$

(b) *Density of cork*

Volume of cork $= 50 \text{ cm}^3$

Mass of cork $= 12.5 \text{ g}$

$$\text{Density of cork} = \frac{12.5}{50} = 0.25 \text{ g/cm}^3$$

The density of a liquid may be found if the mass of a measured volume of a liquid is found. This might be done as follows:

	Example
(i) Find the mass of an empty beaker.	Mass of beaker $= 48.0$ g
(ii) Measure 100 cm^3 of paraffin using a measuring cylinder. Put this in the beaker and find the mass of beaker and the paraffin.	Mass of beaker + paraffin $= 128$ g
(iii) By subtraction, find the mass of the paraffin.	Mass of paraffin $(128 - 48) = 80$ g
(iv) Find the mass of 1 cm^3 of paraffin.	Mass of 1 cm^3 paraffin $= \frac{80}{100} = 0.8 \text{ g/cm}^3$

Finding the density of a gas is rather more complicated and, since the volume of a certain mass of gas changes with temperature and

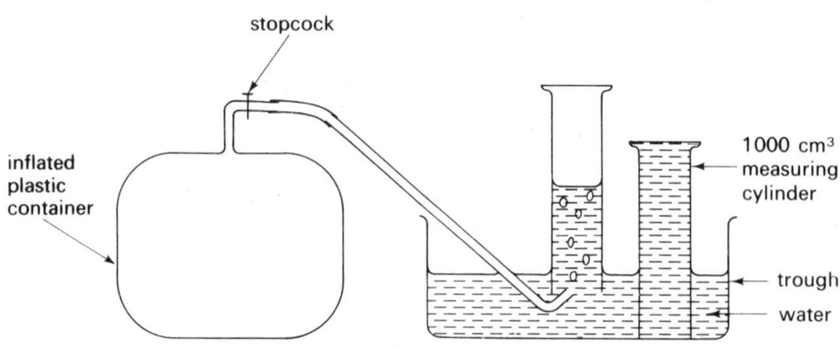

Figure 11

pressure, the density will be different at different combinations of temperature and pressure. Figure 11 shows apparatus which may be used to find the density of air at room temperature and pressure.

A large plastic container fitted with a stopcock is pumped with air until its sides bulge. The mass of the inflated container is found. Air is then released, as shown in the diagram, to fill a number of cylinders so that the volume coming out is measured. This is carried on until no more air comes out. The mass of the container is then found again and, by subtraction, the mass of air which has been released. The volume of the air released has been found so the density of air can be calculated. At room temperature and normal pressure, one litre of air (1000 cm^3) has a mass of about 1.28 g. The density of air is therefore about 0.001 28 g/cm^3.

Self-test

MASS WEIGHT VOLUME DENSITY MENISCUS cm^3 m^3 DISPLACES TEMPERATURE PRESSURE KILOGRAM DIVIDING CYLINDER GRAVITY

Using the words in the list above, copy out and complete the sentences A to J:

A The amount of space anything takes up is called its ______ and may be measured in ______ or ______ .

B A measuring ______ is used to find the ______ of liquids.

C A ______ is the curve formed at the surface of liquids in a tube or beaker.

D The volume of an oddly shaped object may be found by measuring the volume of water which it ______ .

E The volume of a gas may change if there are changes in ______ or ______ .

F How much 'matter' anything contains is called its ______ and this is measured in ______ .

G ______ is the measurement of the force of ______ acting on anything.

H ______ is the ______ of 1 cm^3 or 1 m^3 of a material.

I The ______ of a man is more on Earth than on the moon.

J If 100 g of a material has a volume of 50 cm^3 its ______ is 2 g/cm^3.

Questions

1 Explain fully how you would find the volume in cm^3 of (a) a pint of milk (b) a pebble (c) a table tennis ball.

2 How would you find, as accurately as possible, the volume and the mass of small, steel ball bearings less than 1 mm in diameter? A large number of the bearings, all exactly the same, is available.

3 A certain type of wood has a density of 0.5 g/cm^3. What would be

the mass of a piece of it measuring 100 cm long, 10 cm wide and 2 cm thick?

4 Explain how you would measure the volume of air contained in a blown-up balloon.

5 Describe how you find the density of a liquid such as milk or paraffin.

3 Forces

You use forces every day in one way or another. A push, a pull, a twist, a stretch, a bend are all forces. If a trolley is pushed it begins to move and, if it is already moving, a pull on the handle will cause it to slow down and stop.

Forces change the speed of anything. (If the trolley is at rest and then it is made to move, its speed has been changed.) Forces can also change the direction in which anything is moving and they can change the shape of objects.

If a block of wood is given a push whilst it is resting on a smooth surface, it will slide along the surface and then come to rest. The push is a force applied for a small time. It accelerates the block and then friction, which is also a force, retards (slows down) the motion of the block and brings it back to rest.

If an object is held up and released, it falls downwards (unless it is a hydrogen-filled or a helium-filled balloon.) This is because the Earth exerts a force, the force of gravity, on all objects. It exerts a force on the hydrogen-filled balloon too: the reason that it rises can be understood when FLOTATION has been studied (see chapter 5). All objects exert a gravitational pull, but it is only in the case of very massive bodies, like planets, that this gravitational force is large enough to be measured fairly easily.

The gravitational force does not just pull an object towards the centre of the Earth, it accelerates it. A falling object goes faster and faster. Photographs may be taken of a falling object so that a picture is made at intervals of, say, 1/20th of a second. This will show how far the object has travelled in 1/20th second intervals. Such a photograph is shown in Figure 12. The object travels a greater distance in each 1/20th of a second as it continues to fall. It is accelerating. (Acceleration will be studied in more detail in chapter 6.)

The unit of measurement of force is the NEWTON, (N), named after Sir Isaac Newton who worked out the 'theory of universal gravitation'.

For simple purposes it may be taken that the force of gravity acting on a mass of 1 kilogram at the Earth's surface is 10 newton. From this it follows that a force of 1 newton acts on a mass of 100 g. To hold a mass of 100 g is to feel on it a force of 1 newton.

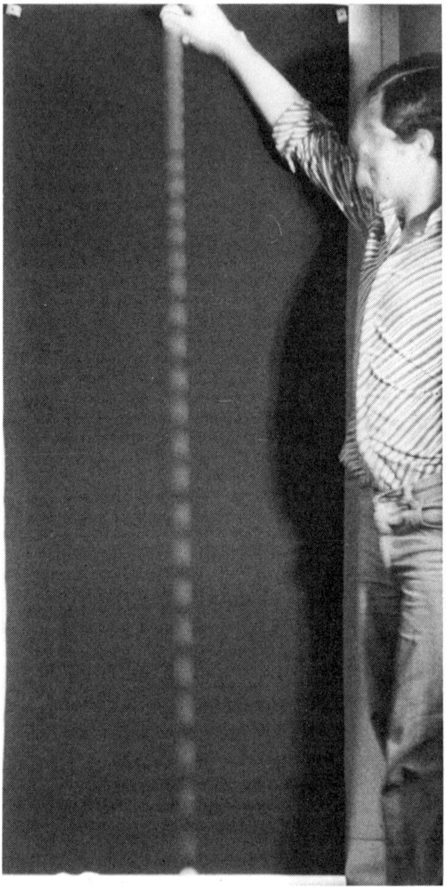

Figure 12

Measuring forces

When a force is applied to a spring, it stretches. Being an elastic material, the wire of the spring goes back to its original shape unless it is stretched beyond its elastic limit.

An experiment to investigate the stretching of a spring may be carried out as follows. A spring is hung against a rule and loads are applied, one at a time; the length of the spring is measured as each load is added. With a steel wire spring, masses of 100 g each, on which the gravitational pull, or weight, is 1 newton, may be used. The length of the spring is recorded each time an additional load of 1 newton is added (see Figure 13).

Table 1 shows an example of the results from such an experiment.

Table 1

Load (N)	0	1	2	3	4	5	6	7	8	9	10
Length (cm)	6	8	10	12	14	16	18	20	24	32	60

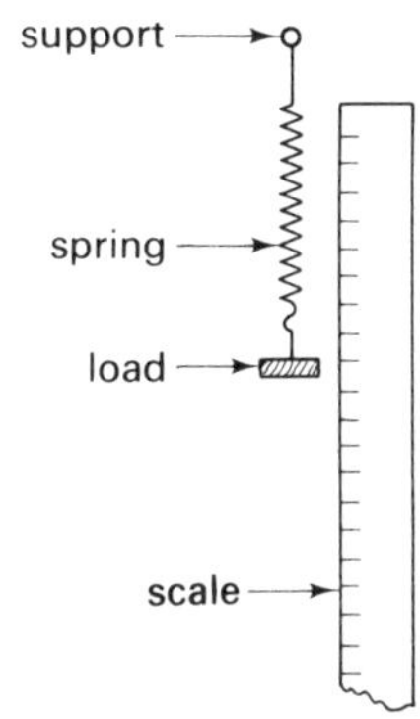

Figure 13

These results are plotted on a graph to show the EXTENSION (the increase in length of the spring) as the load is increased (see Figure 14).

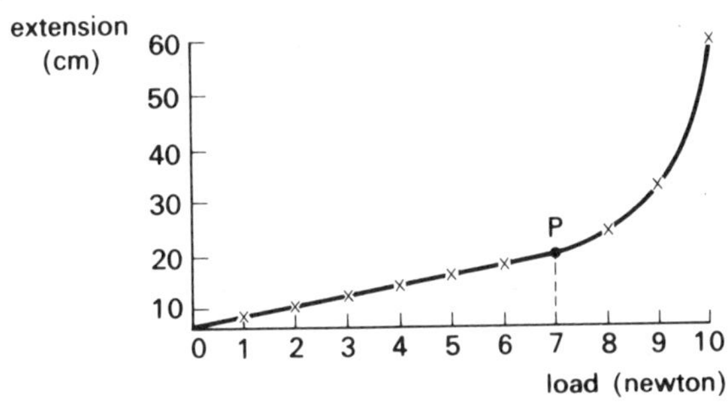

Figure 14

Up to a load of 7 N, the extension of the spring is equal (2 cm) as each load of 1 N is added. Beyond 7 N, the extension increases because the elastic limit (marked on the graph by point P) has been exceeded. The spring does not go back to its original length.

If a spring is used within its elastic limit it can measure forces as pushes and pulls. Figure 15 shows how a force meter may be made using elastic bands. With such a meter the force required to do such things as opening and shutting a door (both at the handle and close to the hinges), lifting a stool or pulling a mass along a bench, may be measured.

In use the plastic tubing is held in the hand, pulls being applied at the cup hook and pushes at the other end of the broomstick. The scale is marked off (CALIBRATED) by applying equal loads to the cup hook and marking where the end of the plastic tube comes against the paper each time.

When a balloon is blown up and then released it flies about the room until all the air is gone. If thin string is passed through two

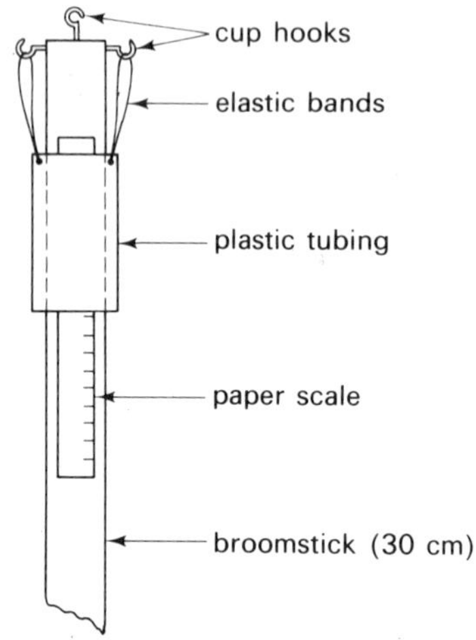

Figure 15

pieces of drinking straw sellotaped to a long balloon and then stretched across a room (see Figure 16), the released balloon can be made to move quite rapidly along the string. Since a force is needed to make the balloon move, it follows that a force must be acting on the balloon when air is coming out of it. The air is made to rush out by the pressure inside the balloon and, when this happens, an equal and opposite force acts on the balloon.

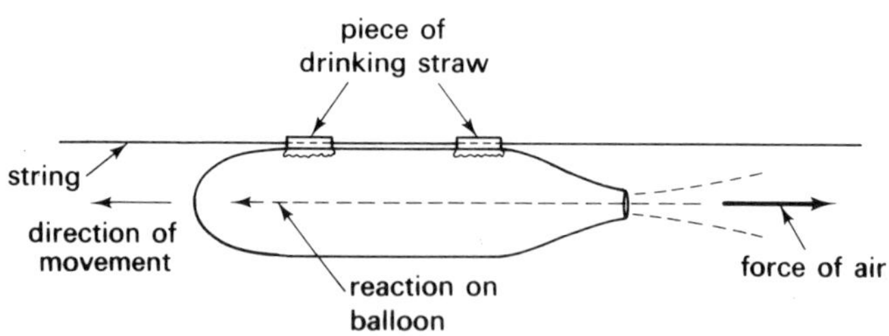

Figure 16

This is the same in the case of rockets and jet engines. Both are driven forward by REACTION, the force acting in the opposite direction to the force exerted by the hot gases. For *every* force there is an equal and opposite reaction. A person trying to push a trolley along on a sheet of ice would probably not succeed. The equal and opposite force from the trolley acting through his feet on to the ice would be greater than the very small frictional force of the ice, and his feet would slip.

The greatest force which a man's muscles is able to exert is often not enough to lift a heavy object and to move, for instance, a screw thread, if he makes use only of his arms and fingers. A man can lift a heavy box by using a lever and he can tighten or loosen a nut if he uses a spanner. When using a lever or a spanner, a man makes use of what is called the MOMENT or the turning effect of a force.

By setting up a balanced beam and using weights (masses on which gravity exerts a known force), moments can be shown at work. In Figure 17 a metre rule is balanced at its centre. A load of 6 N is placed 20 cm from the balance point and then a load of 3 N is placed on the opposite side of the rule and moved about until the rule balances. If the distance from the centre is measured, it is found to be 40 cm. The experiment can be repeated with different loads. Balance is always obtained when each load multiplied by its distance from the point of balance (FULCRUM) gives the same answer.

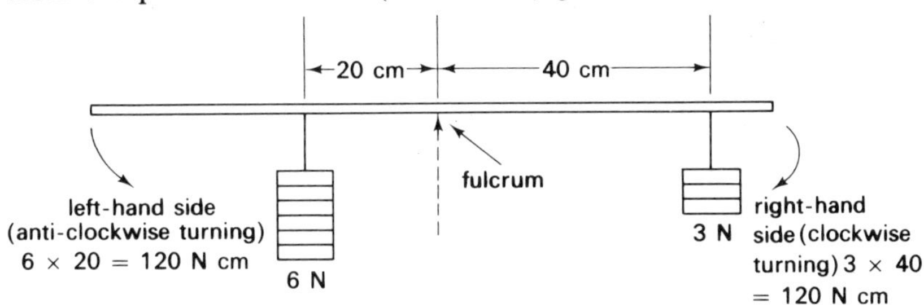

Figure 17

Figure 18 shows how a man might lift up a heavy load using a strong metal bar. Although he will be able to lift it, the force he uses will move through a much greater distance than the load will be lifted.

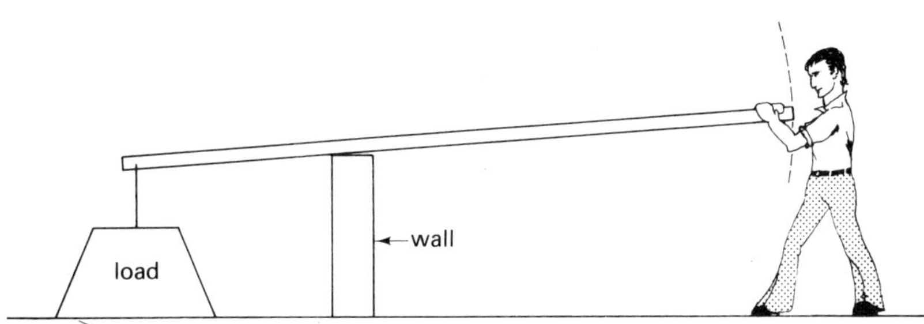

Figure 18

Spanners, scissors, pliers, crowbars and wheelbarrows are all forms of levers based on the PRINCIPLE OF MOMENTS. They are really simple forms of machines. Machines do not give men magical powers of great strength: they are devices which enable small forces to be used through big distances to produce large forces acting through small distances.

When the metre rule shown in Figure 17 was hung up, a thread was tied at its middle point (50 cm). It balanced there because an upward force (the tension in the cord) was equal to the force of gravity acting on the rule (its weight) and that force was applied in exactly the right place. For an object shaped like the one shown in Figure 19, a thread to make the object 'balance' could not be placed in the middle of its length. The force of gravity seems to act in another place which could be found by trial and error. This point is called the centre of gravity or the centre of mass.

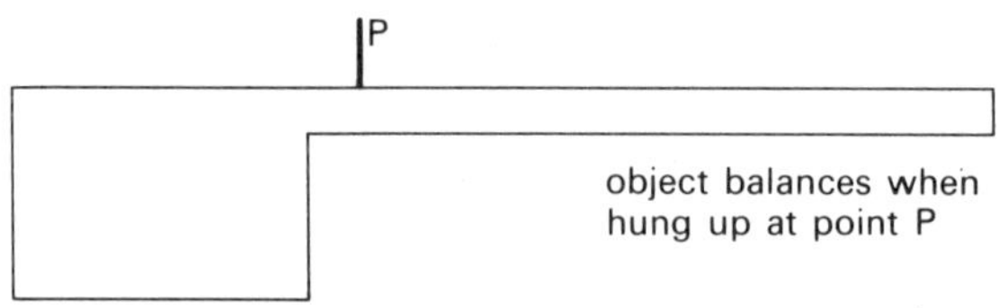

Figure 19

The position of the centre of gravity is very important in designing many things from table lamps to buses and aeroplanes. When it is important that anything should be STABLE – that is, it should not topple over easily when tilted – it should be designed so that the point through which its weight appears to act is as low as possible. If a line going straight down from the centre of gravity falls outside the base of the object, it will topple over. Figure 20 shows this in the case of (a) a block of wood and (b) a block of wood of the same overall size but with a piece of metal fixed to its bottom so that the centre of gravity (CG) is lower; (b) can be tilted much further than (a) before that line falls outside the base.

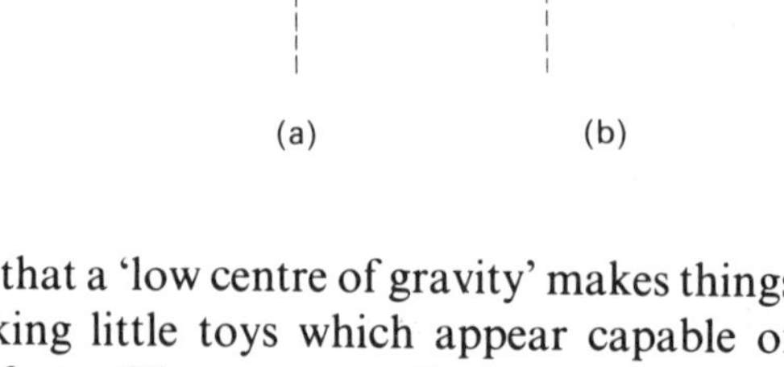

Figure 20

The fact that a 'low centre of gravity' makes things stable is put to use in making little toys which appear capable of extraordinary balancing feats. These can easily be made with such things as knitting needles and polystyrene balls as shown in Figure 21.

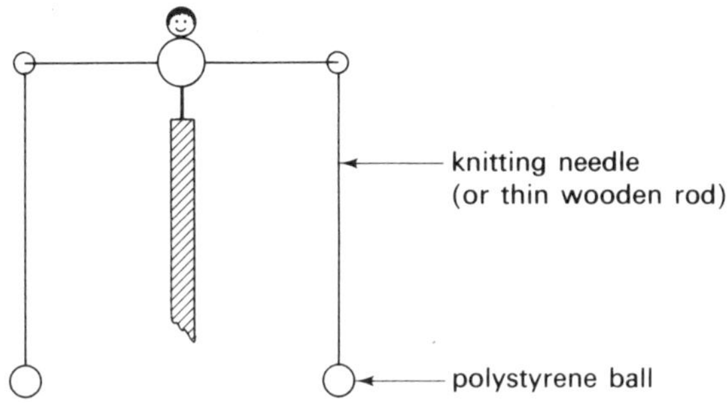

Figure 21

Friction

Figure 22 shows a block of wood resting on a long, smooth plank.

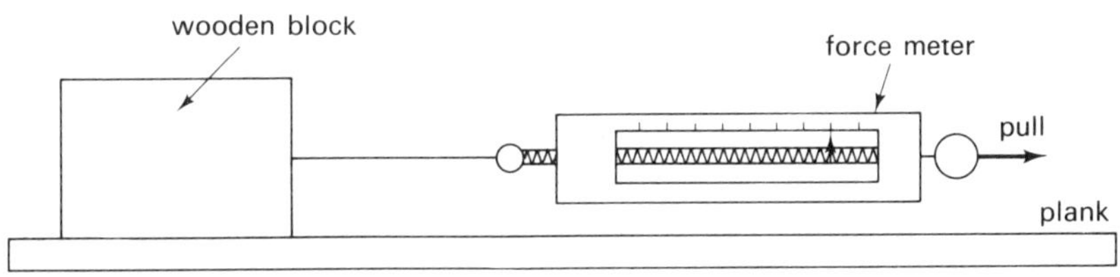

Figure 22

The block is pulled by a force which can be measured in newton by the force meter. In such an experiment it may be found that a force of 3 N is needed to make the block move from rest and, when it is moving, the force needed to keep it moving at an even speed (not accelerating) drops to 2 N. If a force of more than 2 N is now exerted, the block accelerates. Friction is a force met wherever two surfaces are in contact and movement takes place. The smoother the surfaces are, the less the force of friction, and friction may be reduced by using LUBRICANTS like oil. Very small frictional forces can be obtained when the moving object rests on a cushion of air, as in the hovercraft. This means that the force exerted by the engines of the hovercraft is nearly all used to move the hovercraft and not to overcome friction.

In the example shown in Figure 22, the force of friction acting against the motion is, in fact, 2 N; what the force meter reads when the block is moving without accelerating. The reading of 3 N shows what is called the LIMITING STATIC FRICTION; the force which acts against the block being moved from rest.

It would be wrong to think that the force of friction is always a nuisance. Without friction acting on our shoes, we would fall over when attempting to walk and without friction a car's tyres could not 'grip' the road.

Self-test NEWTON REACTION ACCELERATE ELASTIC LIMIT OPPOSITE MASS MOMENT CENTRE OF GRAVITY STABLE MULTIPLYING FORCE FRICTION ACCELERATION

Using the words in the list above, copy out and complete the sentences A to J:

A Pushes, pulls and twists are all examples of ______ .

B ______ is a force which acts when one surface moves against another.

C A force produces or tries to produce ______ .

D Springs generally stretch by equal amounts as equal loads are added until the ______ is reached.

E For every force there is an equal and ______ ______ .

F A plank of wood can be made to balance if it is supported at its ______ .

G A ______ of force has a turning effect.

H A ______ object is one which does not easily fall over if tilted.

I The force of gravity acting on a mass of 1 kg at the Earth's surface is about 10 ______ .

J A moment may be calculated by ______ the force by the distance to the point where it acts.

Questions

1 Describe a simple device for measuring forces and explain how it could be calibrated in newton.

2 A metre rule is balanced at its middle point and a load of 5 N is placed at the 10 cm mark. At what mark on the rule would a load of 10 N be placed to make the rule balance?

3 Explain why friction is thought of as a force. Give one example, from everyday life, of where friction needs to be reduced and one where it needs to be kept as high as possible.

4 Explain why a long spanner is better than a short one when it is necessary to undo a rusted nut from a screw thread.

5 What is meant by 'centre of gravity' (or centre of mass)? How would you find the centre of gravity of a wooden spoon?

4 Pressure

In Figure 23(a), a weight of 10 N is shown placed on a wooden block measuring 10 cm by 10 cm; in Figure 23(b) the same weight is shown placed on a wooden block of the same mass as the other, but measuring only 5 cm by 5 cm in area. Both weights are on a tray of soft clay and it is apparent that (b) sinks further into the clay than (a). Since the force of gravity acting on (a) and (b) is the same, the reason that (b) sinks further into the clay must have something to do with the area of the blocks in contact with the clay. In (a) the area is 100 cm^2 and in (b) it is 25 cm^2. In (a), the force acting over each cm^2 is 10 N/100 cm^2 or 0.1 N/cm^2 and in (b) it is 10 N/25 cm^2 or 0.4 N/cm^2.

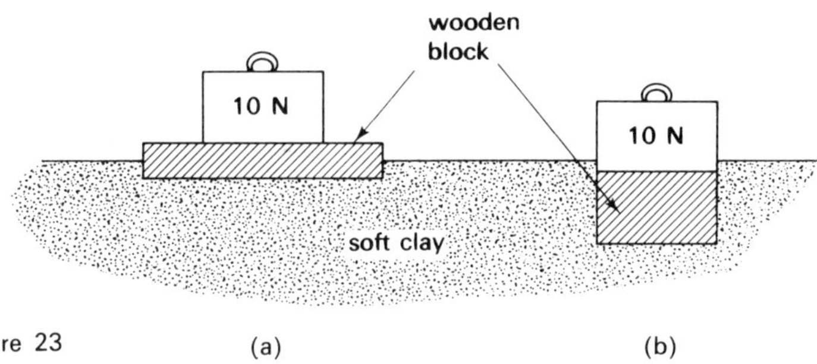

Figure 23

The PRESSURE in (b) is greater than in (a). Pressure may be measured in units of force for each unit of area. The pressure in car tyres is measured in this country in pounds per square inch and, on the continent, in kilograms per square centimetre. In physics, pressure is usually measured in newton per square metre (N/m^2) though, as will be seen later, there are other ways of measuring pressure. A pressure of 1 N/m^2 is called 1 pascal (Pa), the unit of pressure.

For measuring pressure in a gas or a liquid, an instrument called a Bourdon gauge may be used. This works in much the same way as the party 'squeaker', a rolled up paper tube with a coiled spring inside which, when one blows into it, unwinds. Figure 24 shows the Bourdon gauge. The long paper tube of the 'squeaker' is replaced by a metal tube shaped into a curve which straightens out a little when air is forced into it. The small movement moves a 'rack' working against a small cog wheel called a pinion and this, in turn, causes a pointer to move against a dial.

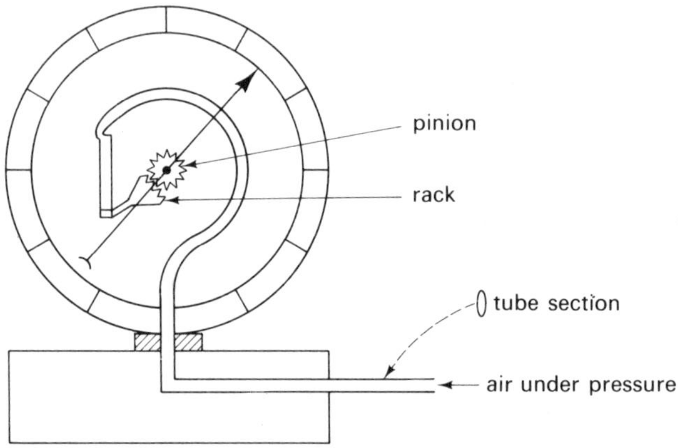

Figure 24 A Bourdon gauge

Another way in which pressure may be measured is by using a 'U'-tube containing a liquid. This device is called a manometer and it may be used for measuring the low pressure under which gas is delivered from laboratory gas taps. Figure 25 shows such a manometer which contains water (a) before it is connected to the gas tap and (b) afterwards. The measurement h cm shown in (b) measures the pressure* of the gas in centimetres of water. If another tube of much wider diameter were to be used, the height would be the same. Water in a manometer is satisfactory for low pressures. The small pressure is only able to 'hold up' a small column of water. If a water manometer were to be used for measuring, say, the pressure you could exert by blowing into a tube, it would have to be very long, about three metres high in fact.

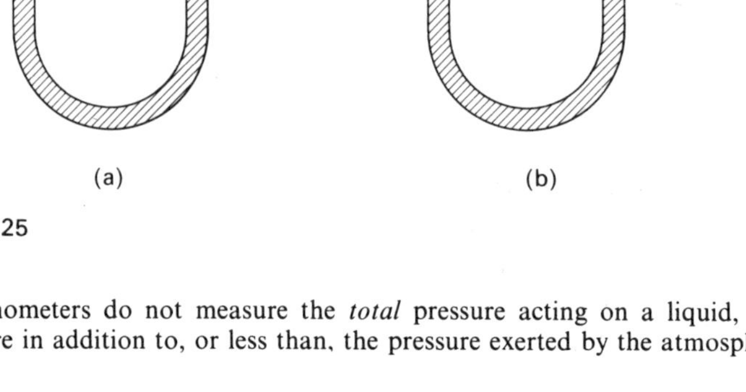

Figure 25

* Manometers do not measure the *total* pressure acting on a liquid, but the pressure in addition to, or less than, the pressure exerted by the atmosphere.

A manometer for measuring higher pressures uses mercury instead of water. Mercury is much denser than water: density of water is 1 g/cm^3; density of mercury is 13.6 g/cm^3. Using a manometer with mercury in it, the pressure would be measured in centimetres of mercury. Doctors frequently use a mercury manometer to measure blood pressure.

The pressure which a liquid exerts depends upon the depth (or height) of the liquid and its density. The deeper a diver descends into water, the greater the pressure on his body.

Figure 26 illustrates a simple experiment which shows that the pressure in a liquid is greater at the bottom than at the top. A tall cylinder has small holes drilled at intervals along its height. Water is run in so that the level is kept the same as water escapes through the holes. The water at the bottom squirts out much further than it does from the holes nearer the surface.

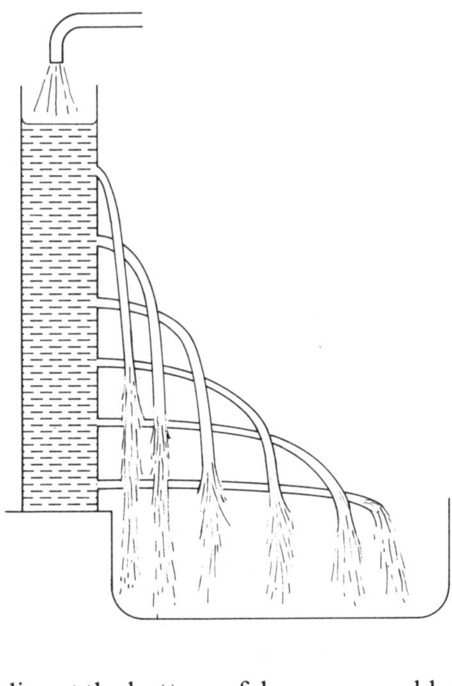

Figure 26

Fish which live at the bottom of deep sea are able to stand the high pressure acting on their bodies. We are somewhat like fish in this respect since we live at the bottom of an ocean of air about 250 miles (400 km) deep. Air has mass and therefore this tall column of air exerts quite a considerable pressure.

Air, or ATMOSPHERIC, pressure is measured with a barometer. A mercury barometer is really a simplified mercury manometer. A strong glass tube about 85 cm long is carefully filled with clean mercury right up to the top, avoiding any trapped air bubbles. A finger is held over the open end of the tube which is then turned upside down and placed below the surface of a small reservoir of mercury. When the finger is removed, some of the mercury, but not all, comes out of the tube into the reservoir.

Mercury, particularly mercury vapour, has harmful effects. The construction of a barometer in the laboratory should not be attempted without careful safety precautions. Figure 27 shows a mercury barometer after it has been set up.

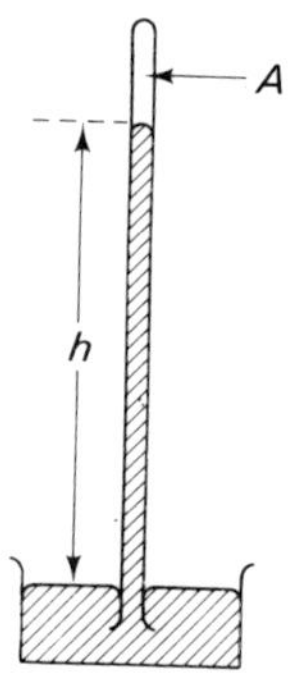

Figure 27

The vertical height h measures the atmospheric pressure in millimetres or centimetres of mercury. Atmospheric pressure varies considerably and is measured by meteorologists because it assists them in making weather forecasts. 'Normal' atmospheric pressure is 760 mm (76 cm) of mercury.

The space A at the top of the tube is nearly a vacuum: it would contain a little mercury vapour.

760 mm of mercury is a pressure which is about equal to 14.7 lb/in^2 (101 325 Pa). Some idea of what that pressure can do may be seen by fitting a one-gallon metal can with a rubber bung and tube and removing air from it with a vacuum pump. The effect of doing this can be seen in Figure 28.

Because mercury is very expensive and because a mercury barometer is a very large instrument, an aneroid barometer is now more frequently used in homes. While not as accurate as a mercury barometer, it is cheaper, more easily read, smaller and less prone to damage. Instead of mercury, thin corrugated metal capsules called ANEROID capsules are used. These have had a little air removed from them and, when pressure on the outside decreases, the capsule expands. When pressure on the outside increases, the capsule is

Figure 28

compressed. The tiny movements of the capsule are made greater by a system of levers, and a pointer against a scale enables atmospheric pressure to be read, generally in centimetres or inches of mercury. (See Figure 29.)

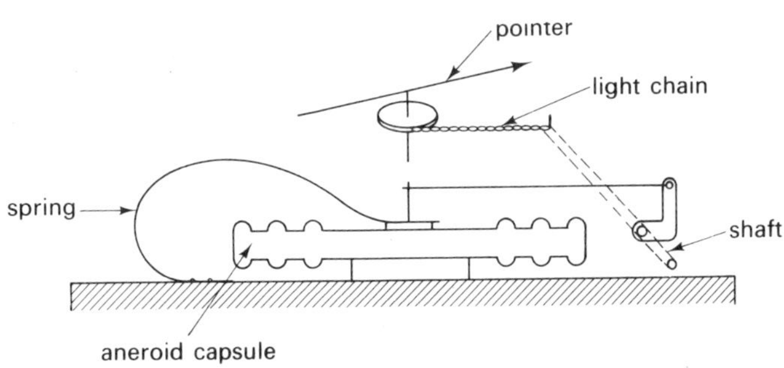

Figure 29

Since liquids, unlike gases, are not compressed when a pressure is applied to them, they are able to TRANSMIT a force from one place to another through tubes. The HYDRAULIC braking system on a car makes use of this principle. The force applied by the driver's foot is transmitted by hydraulic fluid (a light oil) through tubes to the actual brake pads.

A hydraulic lift or press is similarly worked. It may be used to lift very heavy loads with a comparatively small force. The lift in a garage which enables a car to be raised so that work can be done underneath it is often worked hydraulically. The diagram (Figure 30) shows the idea in a very simplified way.

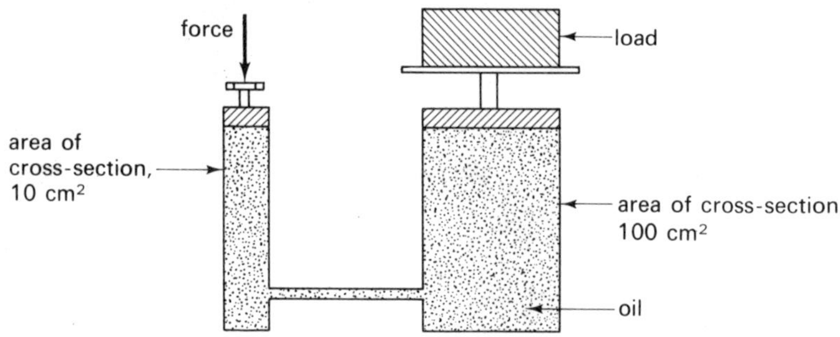

Figure 30

If a force of 10 N is applied to the left cylinder, the pressure (force/area) will be 1 N/cm^2. This pressure is transmitted throughout the fluid. The other cylinder has a much bigger area of cross-section, 100 cm^2, so that a load of 100 N can be supported. (Pressure = force/area = 100 N/100 cm = 1 N/cm^2.)

Self-test

AREA MERCURY FORCE DENSITY BAROMETER
ANEROID MANOMETER BOURDON UNIT
ATMOSPHERE PRESSURE VACUUM N/m^2 (Pa)

Using the words in the list above, copy out and complete the sentences A to J:

A ______ is worked out by finding the force which is acting over a certain ______ .

B Pressure, in physics, is measured in ______ .

C A 'U'-shaped tube containing a liquid is called a ______ and may be used for measuring ______ differences.

D An instrument used to measure the pressure of the atmosphere is called a ______ .

E The pressure in a fluid depends upon the depth and the ______ of the fluid.

F A barometer which uses a thin metal capsule is an ______ barometer.

G A ______ gauge measures ______ by the straightening of a curved metal tube.

H A manometer filled with ______ could measure higher pressure than one filled with water.

I The ______ of the ______ is normally about 760 mm of mercury.

J The space at the top of a mercury barometer is practically a ______.

Questions

1 Describe the way in which a mercury barometer is made. How is it used to measure the pressure of the atmosphere?

2 Explain the difference in the readings of a barometer when it is taken, on the same day, from sea level to the top of a mountain.

3 Explain how the pressure of gas from a laboratory tap might be measured using very simple apparatus.

4 A force of 50 N is applied first to an area of 1 m^2 and then to an area of 0.5 m^2. Work out the pressure in each case.

5 Explain why (a) a hydrogen-filled balloon increases in volume as it rises higher in the air and (b) why submarines cannot safely reach the bottom of the deepest oceans.

5 Flotation

The fact that some things float in water and others do not is fairly obvious. A piece of wood floats in water but a piece of iron sinks. Ships can be made of wood or of iron, steel, fibreglass and other materials and they must all float. Floating depends on more than just the choice of MATERIALS.

Figure 31 shows (a) a beaker of water (b) a sphere (P) of expanded polystyrene floating (c) a sphere (W) of the same volume but made of wood also floating and (d) a sphere (M) of the same volume but made of metal which is not floating. It can also be seen that in (b) very little water is displaced, in (c) a little more water has been displaced and in (d) much more water has been displaced. Since in (d) the metal sphere is completely covered by water, the volume of water displaced must be equal to the volume of the sphere.

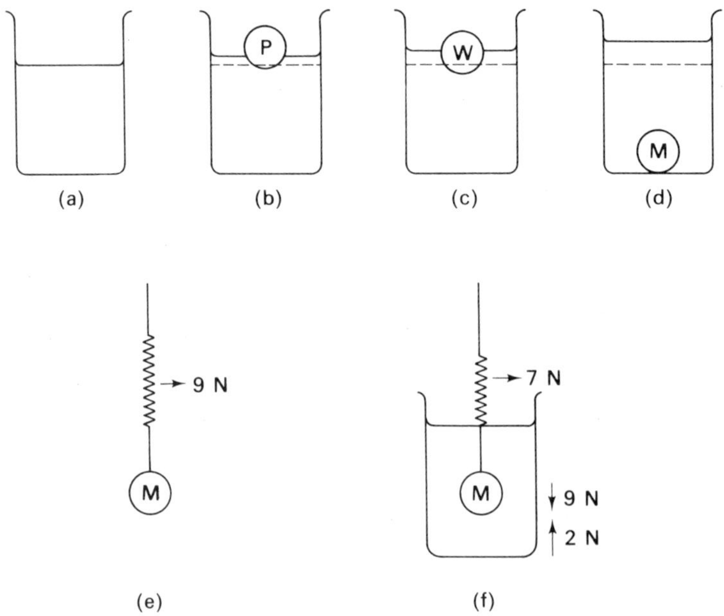

Figure 31

When an object floats it does so because the UPTHRUST is equal to the weight of the object. The upthrust is the force acting upwards exerted by the fluid. In (d), the upthrust is not equal to the weight of the metal sphere and so the sphere sinks. The water does however exert an upward force. If the metal sphere is weighed in air it weighs

9 N as shown in (e): when it is weighed in water (f) it appears to weigh 7 N so the water is pushing upwards with a force of 2N.

The volume of each of the spheres is 200 cm^3. If each of the spheres is placed in a displacement vessel (Figure 32), the volume of water each displaces can be measured. In the case of the polystyrene sphere it is found to be 20 cm^3, the wood displaces 50 cm^3 and the metal sphere 200 cm^3. The density of water is 1 g/cm^3 and, taking the force of gravity on a mass of 1 g to be 0.01 N, the WEIGHT of water displaced by the expanded polystyrene sphere is $20 \times 0.01 = 0.2$ N. A greater volume of water is displaced by the wooden sphere and its weight is found to be 0.5 N. For the metal sphere, the weight of water displaced is 2 N which is equal to the upthrust of the water as measured using the spring balance. When the sphere of polystyrene is weighed, it is found to weight 0.2 N and the wooden sphere is found to weight 0.5 N.

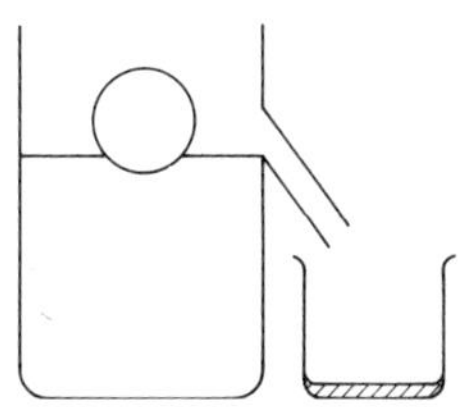

Figure 32

If the WEIGHT of water displaced is equal to the weight of the sphere, it floats because the upthrust of the water is equal to the downthrust of the sphere. Expanded polystyrene and wood are both LESS DENSE than water. Wood is denser than expanded polystyrene so it has to float lower in the water in order to displace a volume of water equal to its own weight. Since the metal is denser than water, even the volume it displaces when completely covered is not sufficient to give an upthrust equal to its weight. How then can a metal object float?

Suppose the metal sphere were to be hammered flat and then made into a box (Figure 33). When placed in the displacement

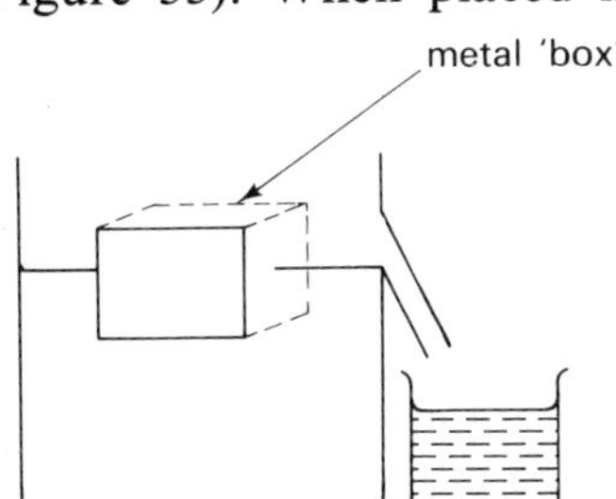

Figure 33

vessel, it displaces 1600 cm^3 weighing 16 N which is equal to its own weight. The upthrust is equal to its own weight, so it floats. This is not due to the fact that 'the box is full of air'. The density of the metal remains the same but it has been made into a shape, a 'body' which is able to displace a volume of water sufficient to equal its own weight.

The principle of upthrust and buoyancy was discovered by Archimedes and is often stated as ARCHIMEDES' PRINCIPLE. *When a body is immersed in a fluid, it experiences an upthrust equal to the weight of the fluid displaced.*

It should be noted that the word 'fluid' is used, and not just water, because the principle applies to all liquids and gases. If the metal sphere is immersed in water and displaces 200 cm^3 of water, the weight of the water is 2 N but, if the sphere is immersed in alcohol, the upthrust is less because alcohol is less dense than water (0.8 g/cm^3). 200 cm^3 of alcohol weighs only 1.6 N. Similarly, if the body were immersed in strong salt water, the upthrust would be more because salt water is denser than fresh water and 200 cm^3 of it would weigh more.

A body floats when the upthrust of the fluid is equal to the downthrust or the body's own weight. In fresh water, a ship will float lower than it would in salt water. It has to displace a volume of water equal to its own weight, and a greater volume of fresh water has to be displaced to equal this weight because fresh water is less dense than salt water. The fact that something floats lower in less dense fluids enables the density of the fluids to be compared and actually measured.

Figure 34 shows a hydrometer. It is a float, usually made of glass with a bulb containing lead shot to make it float upright, and a thin stem marked with divisions. It is shown floating in alcohol and in

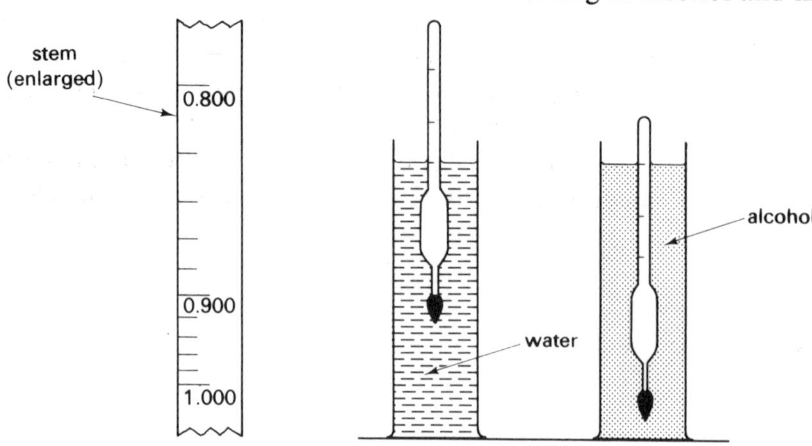

Figure 34

water. The density of water (1 g/cm^3) and of alcohol (0.8 g/cm^3) can be read from the marks on the stem. Hydrometers can be used to find the alcohol content of wines, spirits and beers, the water content of milk and the 'strength' of the acid used in car batteries.

Self-test

WEIGHT UPTHRUST VOLUME DENSITY APPARENT HYDROMETER LOWER HIGHER SALT FRESH LESS MORE DISPLACE VOLUME

Using the words in the list above, copy out and complete the sentences A to H:

A To float in water an object must ______ its own ______ of water.
B A piece of iron appears to weigh less when immersed in water than it does in air due to the ______ exerted by the water.
C Any material which has a ______ of less than 1 g/cm^3 will float in water.
D A ______ is an instrument for measuring the density of liquids.
E A ship will float ______ in fresh water than in salt water because the density of fresh water is ______ than that of salt water.
F A hydrogen-filled balloon will rise in air because the density of hydrogen is ______ than that of air and the ______ of the air displaced is greater than the ______ of the balloon.
G A sheet of metal made into a box will float on water because its ______ has been made ______ .
H A hydrometer will float ______ in the wine which contains the most alcohol.

Questions

1 Describe an experiment which would show that the apparent loss of weight of an iron ball suspended in water is equal to the weight of the water displaced.
2 Explain how a small test tube might be made into a simple hydrometer and how it might be calibrated to read densities.
3 Explain why a heavy object floats in mercury but not in water.
4 A hydrometer will float slightly lower in hot water than in cold. Explain why this is so.
5 A block of wood has a mass of 25 g and a volume of 100 cm^3; a piece of iron has a mass of 80 g and a volume of 10 cm^3. If the iron were fixed to the wood and both placed in water, would they float? Show the calculations used to arrive at an answer.

6 Motion

If a car leaves a certain place, A, and arrives at another place, B, 80 kilometres away, taking exactly two hours for the journey, its AVERAGE speed is 40 kilometres per hour. This figure is obtained by dividing the distance (80 kilometres) by the time (2 hours). The figure obtained is the average speed because it is very unlikely that the car was travelling at the same speed for the whole journey. It would have had to slow down for bends, wait to overtake other cars and possibly stop at traffic lights.

In physics the word VELOCITY is used rather more often than the word 'speed' and it has the special meaning of MOTION IN A STRAIGHT LINE. The car's speed during the journey from A to B would have been made up of many different velocities. Velocity is called a VECTOR quantity. Vectors always have a direction as well as a size. Forces are also vector quantities. Quantities like mass and volume are SCALAR quantities where there is no direction needed or stated.

The time taken for anything to move between two points can be measured with a stop-watch. Stop-watches are usually accurate to 1/5th of a second but accuracy in timing depends on the speed of action of the person using them. Experiments in the laboratory on motion are usually timed with a tape timer. This device is an electrically-operated vibrator, a springy piece of steel strip which vibrates at a rate of exactly 50 times per second. At one end of the strip, a screw taps on to a disc of carbon paper under which there is a paper tape. If the tape is moving, dots are made on the tape and the distance between two dots shows the distance the tape has moved in 1/50th of a second (0.02 s).

Figure 35 shows a trolley which can run down a slope pulling a paper tape with it. Figure 36 shows examples of the tapes which might be obtained from such an experiment.

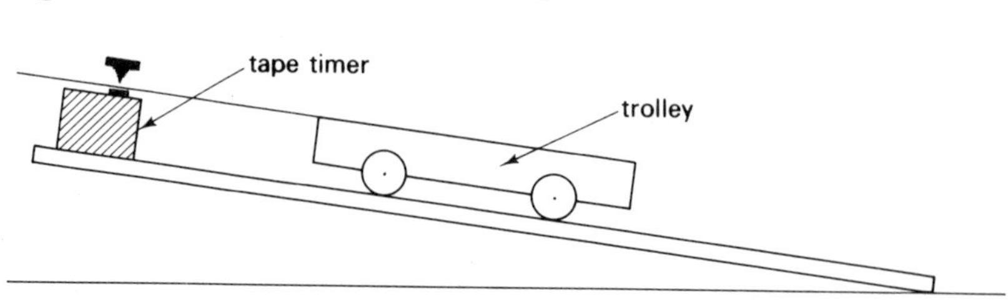

Figure 35

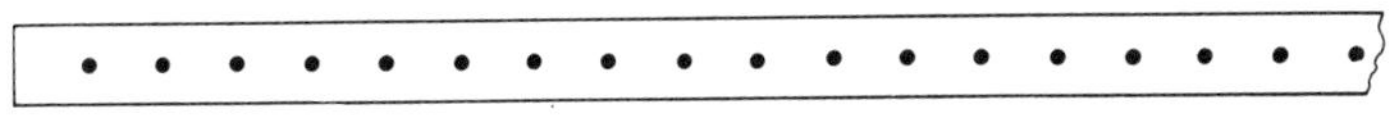
(a) uniform velocity

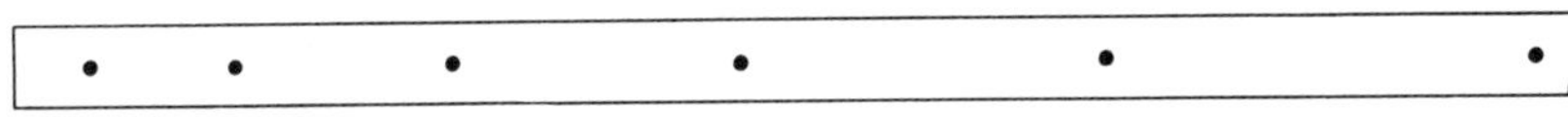
(b) uniform acceleration

Figure 36

In tape (a), the distance between the dots remains the same. This shows what is called UNIFORM VELOCITY: the trolley was travelling equal distances in equal intervals of time. Tape (b) shows that the distance between the dots increases from left to right. This means that the trolley is going faster as time goes on: it is ACCELERATING. If the distances are measured they will show that they increase by 0.5 cm every time. This shows uniform acceleration: motion where the VELOCITY increases by equal amounts in equal intervals of time. Tape (a) would have been obtained with the trolley going down a very slight slope whereas the slope would have been higher to obtain tape (b). If the slope were to be made higher still, the acceleration would be greater and the distance between the dots would increase more. It is the force of gravity which makes the trolley move down the slope. On a slope, the force of gravity acts in two ways: (a) to hold the trolley against the plank and (b) to accelerate it down the plank. As the slope is increased, the force acting down the slope becomes greater than that holding the trolley against it. (See Figure 37.)

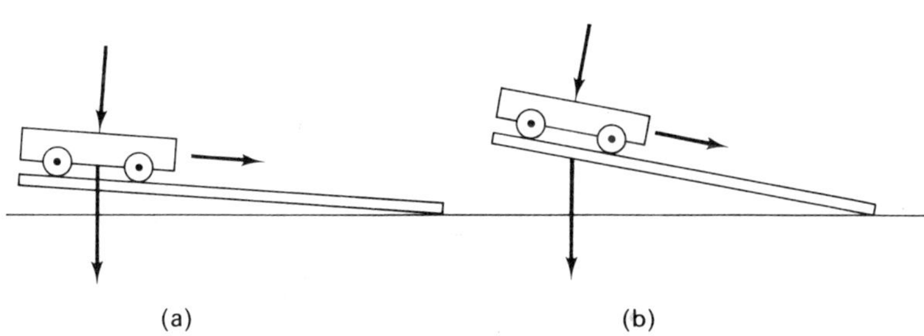

Figure 37

Velocity is usually measured in metres per second (m/s), the distance travelled in a straight line in an interval of time. Acceleration measures how much the velocity changes in an interval of time and is measured in metres per second per second, which is written as m/s^2.

Table 2 shows measurements of the velocity of something moving taken at intervals of one second:

Table 2

Time (s)	0	1	2	3	4	5	6	7	8
Velocity (m/s)	0	4	8	12	16	20	24	28	32

According to these measurements, the velocity increases by 4 m/s every second. This is an example of UNIFORM ACCELERATION of 4 m/s^2.

Figure 37 showed that as the slope was increased, the trolley accelerated more since a greater part of the force of gravity acted down the slope; but when the slope was very slight it was possible for the trolley to go down the slope at a UNIFORM VELOCITY. This requires explanation. If anything is allowed to 'fall freely', the force of gravity makes it accelerate. An example of free fall photographed at short intervals of time is shown in Figure 12. Measurements made from such photographs show that the acceleration due to gravity is approximately 10 m/s^2 or, more accurately, 9.81 m/s^2. This acceleration is the same for all objects, whatever their mass. A mass of 1 kg accelerates at the same rate as a mass of 2 kg. Dropped from a height at the same time, both masses would reach the ground together.

It is obvious that a lead weight and a small piece of paper released together do *not* reach the ground at the same time. Air through which they fall offers resistance which affects the paper more. If a coin and a small piece of paper are placed in a long glass tube from which all air has been removed with a vacuum pump and the tube is then turned upside down, the paper and the coin reach the other end of the tube at the same time.

A similar experiment of free fall can be done with a ball-bearing falling through oil. The ball falls quite slowly and its velocity can be measured using a stop-watch and measurements of distance marked on the tube. The results of such an experiment are shown in the diagram (Figure 38).

The ball-bearing accelerates at first until it reaches point A and then continues downwards with uniform velocity. The same happens to sky divers who jump from an aircraft and do not open their parachutes until they are quite close to the ground. At first they accelerate but, when they have reached a speed of about 160 m.p.h. (255 km/h), their velocity does not increase. This is because the force

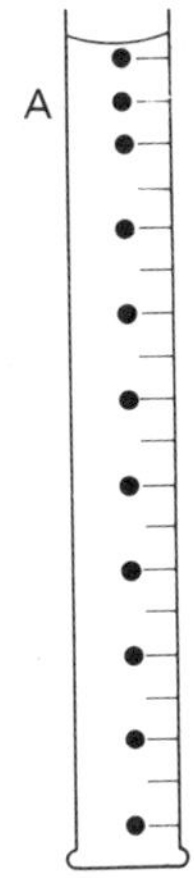

Figure 38

of air resistance has become equal to the force of gravity. In the tube of oil, its resistance soon becomes equal to the force of gravity acting on the ball-bearing.

This also explains why the tape obtained with the trolley on a very slight slope (Figure 37(a)) shows uniform velocity. The force of friction acting on the trolley is equal to the part of the force of gravity acting down the slope.

A trolley can be set up on a small slope so that it will roll down with uniform velocity. A slope set at just the right angle is said to be FRICTION COMPENSATED. When this adjustment has been made, the effect of different forces on the trolley can be found.

Figure 39 shows a trolley in the friction compensated position being pulled down the slope with a force applied by a very thin elastic cord stretched the length of the trolley. This is much the same as applying the force to a trolley resting on a surface where there is *no* friction, something impossible to obtain. By using the tape timer, the motion of the trolley can be recorded. It accelerates: the force causes the acceleration. The force can be doubled by using two elastic cords side by side. The MASS on which the force acts can be increased by stacking one trolley on top of another.

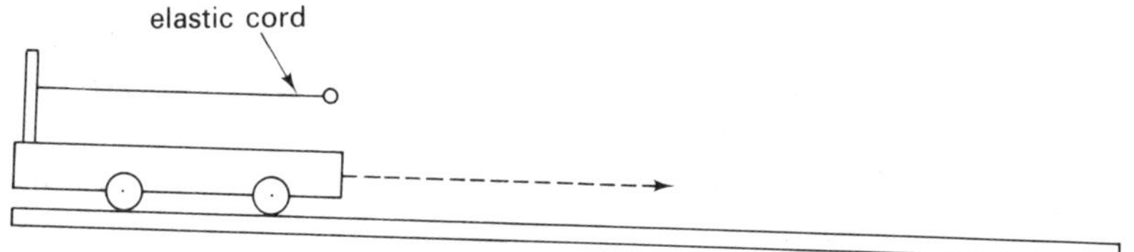

Figure 39

By such experiments, it is found that if one force acts on one trolley it produces an acceleration; if the force is doubled, twice the acceleration is produced. If one force is applied to two trolleys, half the acceleration is produced. (See Figure 40.)

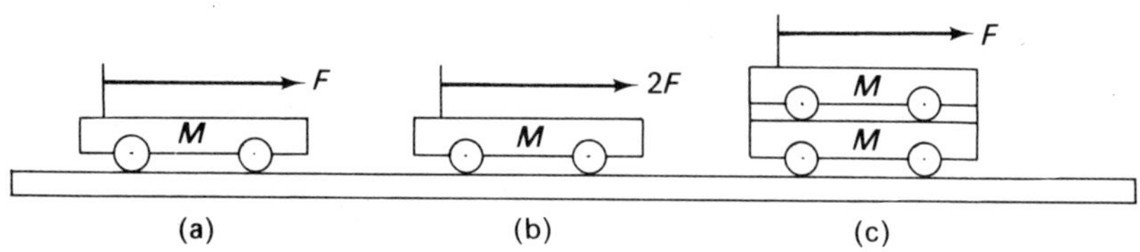

Figure 40 (a) A single force acting on one mass produces an acceleration of a m/s². (b) Two forces acting on one mass produce an acceleration of $2a$ m/s². (c) One force acting on two masses produces an acceleration of $\frac{1}{2}a$ m/s².

From measurements like these it is possible to arrive at an important connection between mass, force and acceleration which is usually written:

$$f = m\,a$$

or force (newton) is equal to mass (kilogram) × acceleration (m/s²). From this it follows that:

$$\text{acceleration} = \text{force divided by mass}$$

For example, if a force of 5 newton acted on a mass of 2.5 kg, the acceleration would be (5/2.5) or 2 m/s². This would be the case if there were no friction opposing the movement. Generally, there will be friction acting against the force accelerating the body, reducing that force and making the acceleration less.

The newton, the unit in which force is measured, is: *that force which, acting on a mass of* 1 kg, *will produce an acceleration of* 1 m/s².

Acceleration, which must include its opposite RETARDATION (slowing down), will always happen if the total of the forces acting on anything is UNBALANCED, which means that, if all the forces are taken into consideration, there will be more force acting in one direction than in another. If anything is at rest, say on a table, the force of gravity acting downward is balanced by the force exerted by the table upwards. If a force of 5 N is applied to the left and a force of 2 N to the right, the object will be accelerated to the left by a force of 3 N (see Figure 41).

If the forces acting on an object are BALANCED one of two things can happen: it can remain at rest or it can be in UNIFORM MOTION IN A STRAIGHT LINE. The first is fairly obvious as in the case mentioned where an object rests on a table. The second is a little

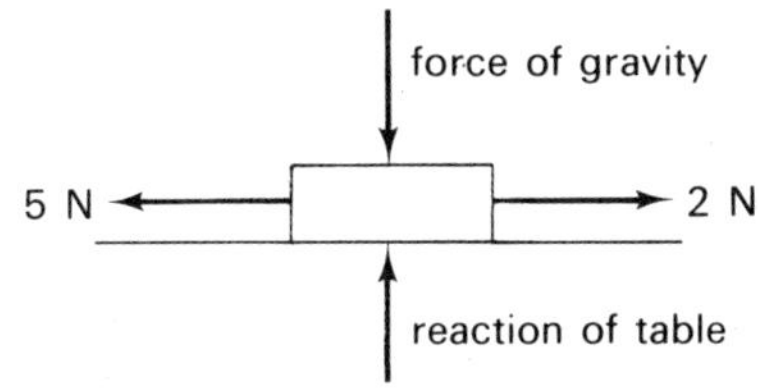

Figure 41

more difficult to understand, but consider, (i) the sky diver who reaches a speed of fall which does not change until he opens his parachute, (ii) the ball-bearing falling through the oil and (iii) the trolley on the slight slope which is friction compensated. In all these cases the force of (i) air resistance, (ii) oil resistance and (iii) friction of the plank on the trolley wheels becomes exactly equal to the force due to gravity and the result is that the motion continues without a change in velocity.

When a car starts from rest, its engine exerts a force to accelerate it. It has to overcome the friction of the road and the air and also to accelerate the mass of the car. When it has reached a 'cruising speed', say 40 m.p.h. (65 km/h), the engine no longer has to accelerate the mass of the car, it simply has to exert a force to overcome the friction of road and air. When these forces are balanced the car will go in uniform motion in a straight line.

Self-test

SPEED VELOCITY VECTOR ACCELERATION
UNIFORM DISTANCE TIME AVERAGE
FORCE MASS BALANCED UNBALANCED m/s m/s^2

Using the words in the list above, copy out and complete the sentences A to J:

A The ______ speed of a car between two places may be worked out by dividing the ______ travelled by the ______ taken.

B Velocity is a ______ quantity which means that the direction must be given.

C ______ velocity is where equal distances are covered in equal intervals of ______ .

D Uniform ______ is where velocity changes by equal amounts in equal intervals of time.

E Velocity is measured in ______ and acceleration in ______ .

F The acceleration of an object depends upon its ______ and the size of the ______ acting on it.

G If the forces acting on an object are ______ it will remain at rest or in ______ velocity in a straight line.

H The ______ due to gravity is approximately 10 ______ .

I When a car is travelling at a uniform velocity, the ______ exerted by the engine and the forces of road friction and air resistance are ______ .

J Force (N) is equal to mass (kg) multiplied by ______ (m/s^2).

Questions

1 A car travels a distance of 100 kilometres in 2 hours 30 minutes. Calculate its average speed and explain why only the *average* speed can be calculated.

2 Explain how a tape timer is used to measure the motion of a small trolley. Draw diagrams to show what the tapes would look like for a trolley in (i) uniform velocity and (ii) uniform acceleration.

3 Calculate the acceleration of a mass of 5 kg pulled by a force of 25 N if there is a steady frictional force of 5 N.

4 Explain why a free fall parachutist reaches a downward velocity of about 160 m.p.h. (255 km/h) which does not change until he opens his parachute.

5 A trolley is travelling at 4 m/s. (i) At this velocity, how far would it have gone in 10 s? (ii) If the trolley accelerated uniformly at 2 m/s^2, what would its velocity be after 5 s?

7 Energy

Energy is one of the most important things in life. It is that which enables us to do WORK. Work in physics, means such things as changing the position of an object by lifting it up or changing its speed, raising the temperature of something, changing the shape of an object or a material and providing light.

We need energy to move about and to do the activities of a normal day. We obtain the energy necessary for this from food and, more particularly, from the starch and sugars of our diet, the CARBOHYDRATES, which the body uses as 'fuel' in much the same way as a steam engine might burn coal or oil.

Figure 42 shows a man doing a simple piece of work. He is pulling on a rope over a pulley to raise a load. The fact that he uses some energy in doing this will be felt by the man. When he has raised the load to the top he might fasten the end of the rope he has been pulling to a box. Provided that the box is lighter than the load, the load would then run down and pull the box up. To do this the load must have some energy. This energy has actually been given to the load by the man: he has TRANSFERRED some energy. When the load is at the top of the distance it has been pulled up, it has gained POTENTIAL energy: it has the ability to do some work but it is not yet using this ability. It is only when the load is moving downwards that it is doing work in pulling the box up. When the load is moving, it has KINETIC energy – energy due to motion.

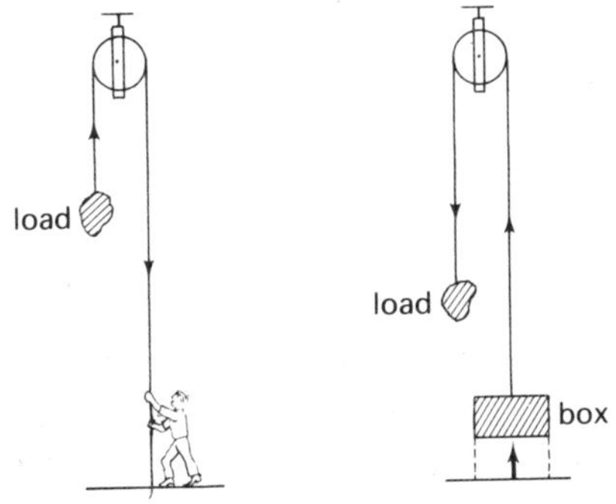

Figure 42

Figure 43 shows another example of potential and kinetic energy. The stretched bow (a) has potential energy: when the bow is released (b), that energy becomes kinetic and is transferred to the arrow.

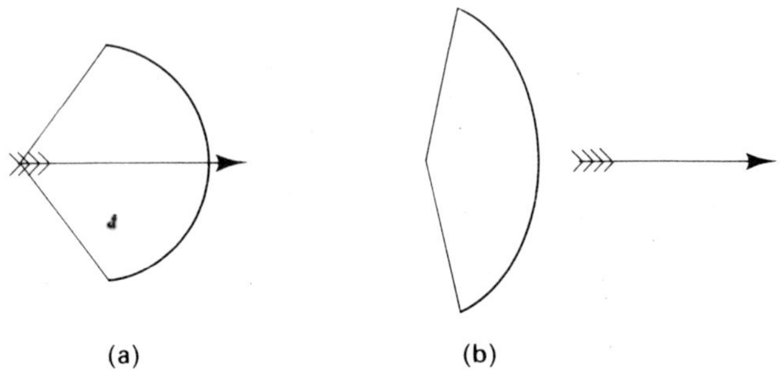

Figure 43

The diagram (Figure 44) shows the various forms which energy can take. Potential and kinetic energy can be grouped together as MECHANICAL energy. Sound energy is a special form of kinetic energy since it is obtained from the movement of molecules (see chapter 14). Chemical energy is a special form of potential energy, the energy 'locked up' in fuels such as coal and oil which is released as heat when the fuel is burnt. Nuclear energy is produced when particles are thrown out from radioactive atoms of materials like uranium and plutonium which are used in atomic piles, yielding great heat.

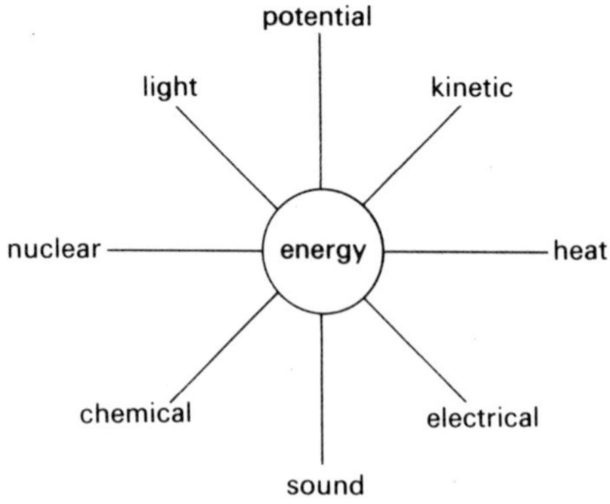

Figure 44

Whenever work is done, one form of energy is changed into another or transferred from one place to another. When this happens, there is no 'loss' of energy: it can all be accounted for in one way or another. This is known as CONSERVATION of energy and is an important principle of physics. We speak of loss of energy when we mean that, for example, only a little of the energy contained in petrol is transferred into kinetic energy in a car engine whilst a great deal escapes into the atmosphere as heat (and sound to a lesser extent).

It is of great importance that the energy we obtain from fuels like oil and coal is used with the least possible waste. Oil and coal are fossil fuels formed millions of years ago from the remains of animals and plants. There is only a limited amount of these fuels contained in the Earth's surface and once this is used up it can never be replaced. Only quite recently has it been realised that our supply of this form of energy is running low. There is much work now being done to see how other forms of energy, obtained from the wind, tides and the sun, might be used for producing the electrical energy we need.

The chain of energy changes which takes place in any particular situation can be worked out; examples are shown in Figures 45, 46 and 47.

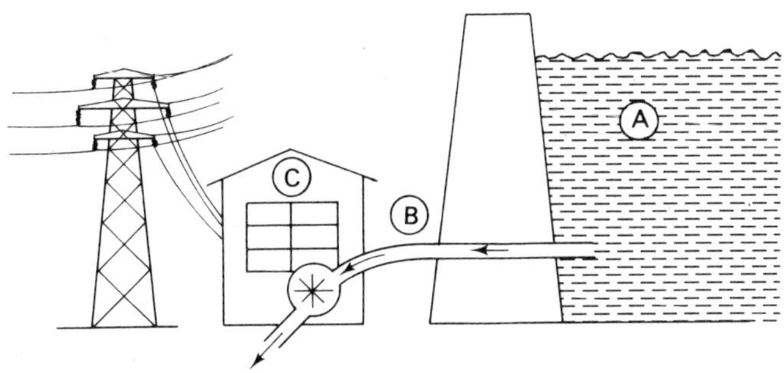

Figure 45 At A the water behind the dam has potential energy. At B the energy produced by the water turning the water turbine is kinetic energy. At C the kinetic energy of the turbine is changed into electrical energy by the dynamo.

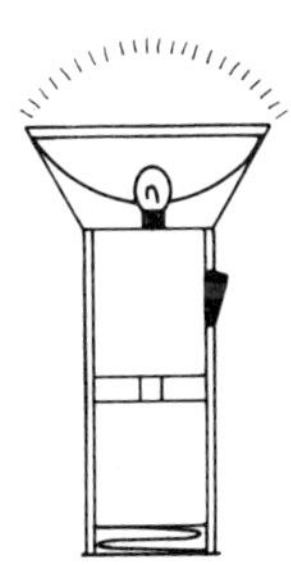

Figure 46 A flashlamp. Chemical energy of the battery is changed to electrical energy which, in the lamp, is changed to heat and light energy. (Light comes from the white hot filament of the lamp.)

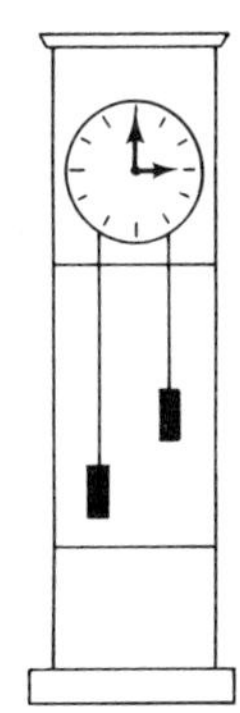

Figure 47 A grandfather clock. Potential energy of the weights is changed to kinetic energy as they slowly descend, turning the hands of the clock, and is changed into sound energy by the chimes.

Measuring energy

All energy is measured in a unit called the joule (J). This is the energy transferred when a force of one newton moves through a distance of one metre.

It takes a force of 1 N to lift a mass of 0.1 kg at the surface of the Earth. If the mass is raised 1 metre it will gain 1 joule of potential energy.

Heat energy is also measured in joules and it may be difficult to see how a unit which involves a force moving through a distance can be used to measure heat. Kinetic and potential energy can be converted into heat, however. If a piece of lead is hammered on a block of wood it becomes flattened as a result but it will also be found to have become quite warm in the process. The potential and kinetic energy of the hammer blows have been converted into heat energy. Kinetic energy is converted into electrical energy by a dynamo and electrical energy is readily converted into heat energy by an electric fire, iron or kettle. One unit of energy, the joule, is therefore all that is needed to measure all forms of energy.

Power

If a mass of 10 kg were to be raised to a height of 5 m, the energy transferred would be as follows:

100 newton through 5 metre, that is 500 J

(100 newton is needed to raise a mass of 10 kg).

The mass might be raised slowly, taking 20 seconds, or it might be raised more quickly, taking only 5 seconds. In both cases, the same amount of energy (500 J) is transferred. In the second case the energy transfer, or work, is done more quickly. If the work were done by a motor of some sort, the motor which could raise the load most rapidly would be the one called the 'most powerful'.

Power has the special meaning in physics of RATE OF ENERGY TRANSFER and is measured in watt (W). One watt is a rate of energy transfer of one joule per second. A one-hundred watt electric lamp is transferring electrical energy into heat and light energy at the rate of 100 joule per second:

Transferring 500 J in 20 seconds represents 500/20 or 25 watt.
Transferring 500 J in 5 seconds represents 500/5 or 100 watt.

Your own body power can be worked out. Find your weight, either by using a balance calibrated in newton or by changing kilogram readings by multiplying by ten. Find the height of a flight of stairs in metre and then, timed by a stop-watch, find how quickly you can get to the top of the stairs. The energy transferred will be your weight (newton) × height (metre). The power will be this amount of energy divided by the time taken (second).

The power of car engines in this country is expressed in HORSE POWER. This comes from the use of the units 'foot' and 'pound' and from experiments by James Watt to find the power of 'the average horse'. Do not forget that, before the invention of the steam engine, the horse was man's main source of 'power'. 1 horse power is equal to 746 watt. For larger power measurement, the unit KILOWATT, equal to 1000 watt, is used.

Self-test

ENERGY POTENTIAL KINETIC NEWTON JOULE POWER SECOND RATE HEAT TRANSFERRED FUEL WATT KILOWATT CONSERVED WORK POSITION MOTION

Using the words in the list above, copy out and complete the sentences A to J:

A Heat, light and electricity are all forms of ______ .
B ______ energy is due to motion: ______ energy is due to the ______ or state of an object.
C In a steam engine, ______ energy is obtained by burning a ______ and this energy is then changed to ______ energy.
D Energy is not actually lost but is changed from one form into another: energy is said to be ______ .
E Energy is measured in ______.
F If a force of 5 N moves through a distance of 2 m the energy transferred will be 10 ______.
G The rate at which energy is transferred is called the ______ and is measured in ______ .
H Power can be worked out by finding how many ______ of energy are transferred in one ______ .
I ______ is done when one form of energy is transferred into another.
J Energy transfer of 1000 joules per second is a power of one ______ .

Questions

1 What energy changes occur in the following:
(i) a bicycle dynamo lighting a lamp?
(ii) a rubber 'motor' of a model aeroplane wound up and then released to turn the propeller?
(iii) wind turning a windmill to pump water into a storage tower?

2 Describe an experiment which could be carried out to find the power of a model electric motor.

3 A load of 100 N is raised 20 m in 50 seconds. Calculate (i) the gain in potential energy and (ii) the power developed.

4 Explain why a petrol engine is wasteful in the way it uses the energy contained in petrol.

5 Describe an experiment which would show how kinetic energy can be transferred as heat energy.

8 Machines

Machines are not necessarily complicated although many, such as those used in factories, are. Even these complicated machines are built up from fairly simple units which man has known about and used for many centuries. A machine is really a device which enables a person to lift or move loads which he would not be able to do using only the fairly small forces which can be exerted by the muscles. A car is a very heavy object but, every day, men and women lift a car's wheels off the ground in order to deal with a puncture, using very little effort but using also a simple machine called a screw jack.

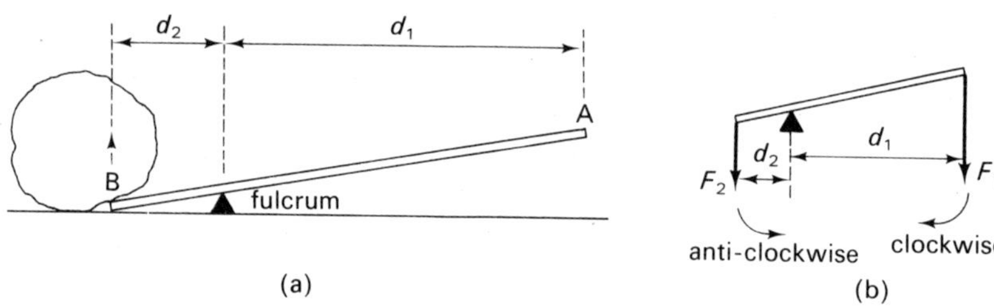

Figure 48

One of the simplest of machines is the lever (Figure 48). It is shown (a) being used to shift a heavy boulder. The lever uses the PRINCIPLE OF MOMENTS, the 'see-saw' principle, mentioned in chapter 3. A small force exerted at A, downwards, will cause a bigger force to be exerted at B, upwards. A force exerted downwards at A would cause the lever to turn about the fulcrum in a clockwise direction. A force exerted downwards at B would cause the lever to turn similarly in an anticlockwise direction. These 'turning effects' are the moments of force and they are found by multiplying the force by the distance *perpendicular* to the fulcrum. When the load (the boulder) is just raised (b), the clockwise and anticlockwise moments of force are equal, that is:

$$F_1 \times d_1 = F_2 \times d_2$$

If the force at A were 100 N and the distance d_1 were 2 m, then the clockwise moment of force would be:

$$100 \times 2 = 200 \text{ N m}$$

If the distance d_2 were 0.5 m, the anticlockwise moment of force would be:

$$F_2 \times 0.5 \text{ N m}$$

Since the moments of force are equal:

$$F_2 \times 0.5 = 200 \text{ N m}$$

$$F_2 = \frac{200}{0.5} \text{ N}$$

$$F_2 = 400 \text{ N}$$

In this way, a force of 100 N has been used to produce a force of 400 N. The machine gives a MECHANICAL ADVANTAGE of 4 which is found by dividing the load (400 N) by the effort (100 N). It will be seen that the effort has to move quite a long way in order to make the load move a smaller distance.

Figure 49 shows a lever used in a different way. The fulcrum is right at the end of the bar, the effort at A is in an upward direction and the load also is lifted upward. If the force at A were 20 N and the distance d_1 were 2 metres, the anticlockwise moment would be 40 newton metre. If the distance d_2 were 0.5 metres, the force acting upwards would be 80 N. There would again be a mechanical advantage of 4.

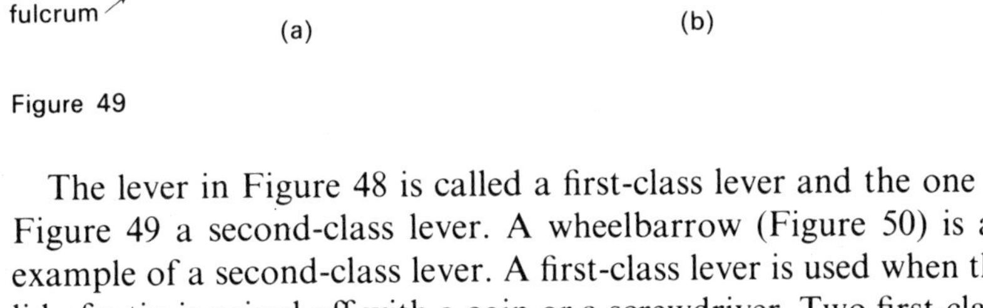

Figure 49

The lever in Figure 48 is called a first-class lever and the one in Figure 49 a second-class lever. A wheelbarrow (Figure 50) is an example of a second-class lever. A first-class lever is used when the lid of a tin is prised off with a coin or a screwdriver. Two first-class levers put together make a pair of pliers; two second-class levers put together make nutcrackers (Figure 51).

To get a heavy barrel into the back of a lorry, a slope, up which the barrel is rolled, is often used. This is a simple machine called an inclined plane. Figure 52 shows a 'barrel' weighing 100 N pushed

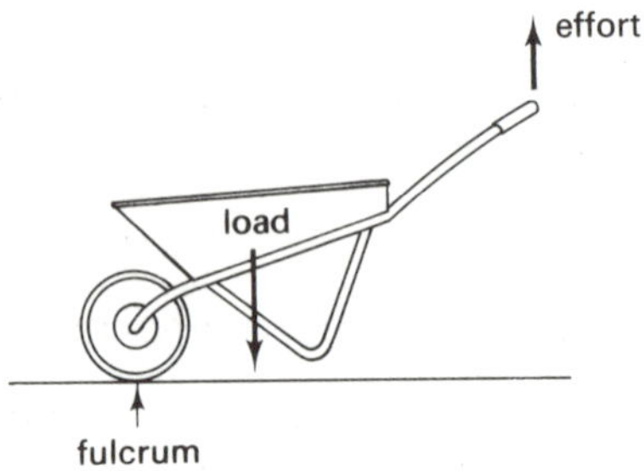

Figure 50

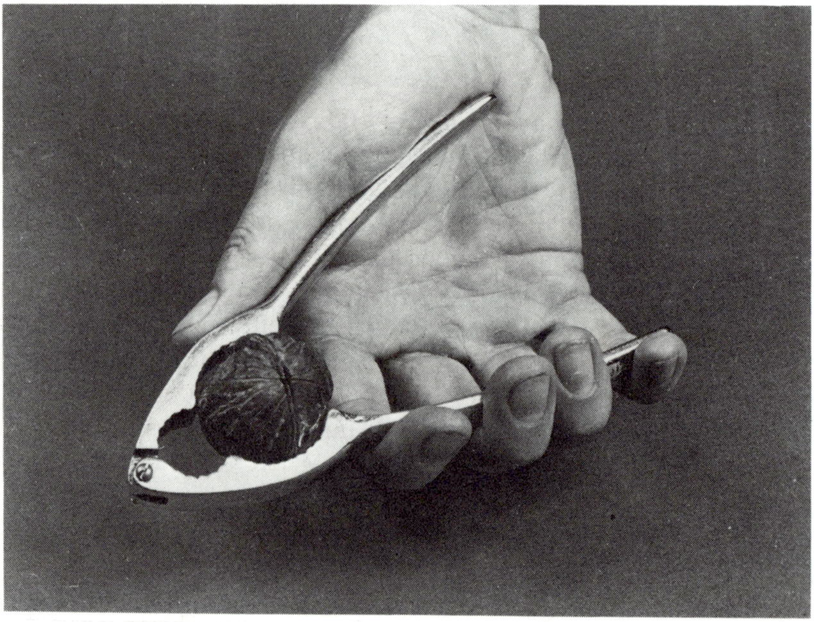

Figure 51

with a force of 30 N up a slope. The barrel moves 5 m along the slope and rises 1 m as it does so. The machine gives a mechanical advantage because a force of 30 N moves a load of 100 N: the mechanical advantage is (100/30) or 3.3. The energy transferred in pushing the barrel up the slope is calculated by multiplying the force by the distance moved:

$$30\ \text{N} \times 5\ \text{m} = 150\ \text{J}.$$

The potential energy gained by the load is

$$100\ \text{N} \times 1\ \text{m} = 100\ \text{J}.$$

So no energy has been 'saved' by the machine, in fact, some has been 'wasted'. The difference of 50 J of energy would be accounted for in overcoming the friction on the slope. No machine will ever give the same energy output as the energy put in because of friction and other effects. The efficiency of the inclined plane in Figure 52 can be calculated:

$$\frac{\text{energy output}}{\text{energy input}} \times 100 = \frac{100}{150} \times 100 = 66\,\%$$

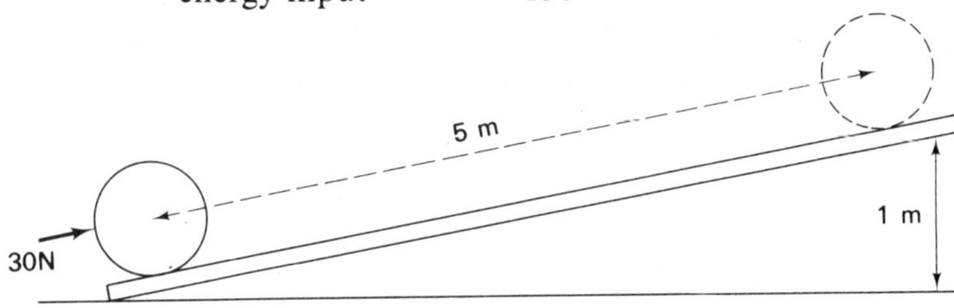

Figure 52

Another point should be noticed about the inclined plane: the EFFORT moves 5 m up the plane while the LOAD rises a height of 1 metre. This is really the secret of most machines – a small force moving a long distance produces a bigger force which moves through a smaller distance. The comparison between the distance the effort moves and the distance the load moves is called the VELOCITY RATIO of the machine and is calculated as follows:

$$\frac{\text{distance moved by effort}}{\text{distance moved by load}}$$

where the distances are measured over the same time. In the case of the inclined plane, the velocity ratio is:

$$\frac{5\text{ m}}{1\text{ m}} = 5$$

A winch, or a wheel and axle, is a simple machine put to all sorts of uses. In the example (Figure 53) it is used at the head of a well. The

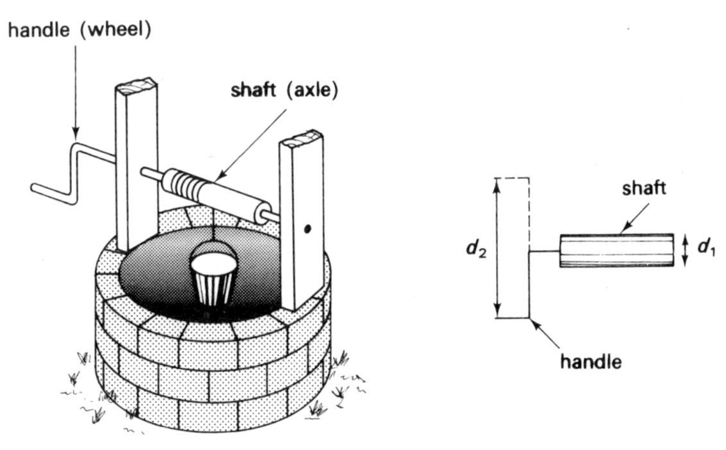

Figure 53

effort applied at the handle moves through a much bigger distance than the rope around the axle. The velocity ratio is found by comparing the diameter of the two circles, d_2/d_1.

Pulley systems also provide a mechanical advantage. Their velocity ratios depend upon the number of pulleys used. Two pulleys can give a velocity ratio of 2; three pulleys a velocity ratio of 3 and so on. Three examples of pulley systems are shown in Figure 54.

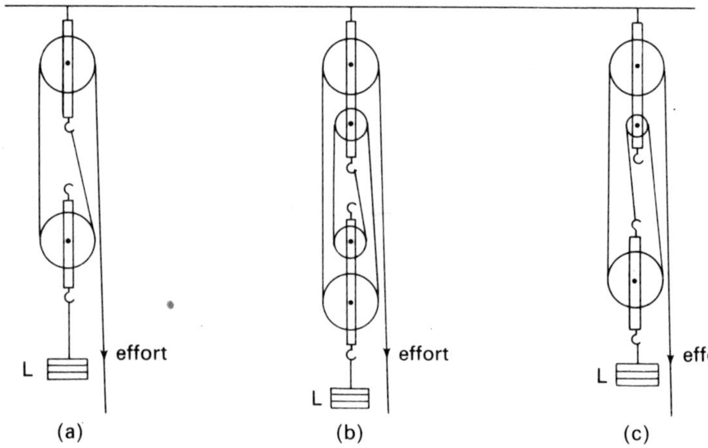

Figure 54

Note: in (b) and (c) the double pulleys are actually side by side and not as illustrated

A screw has a spiral groove cut into it. The number of grooves that occur in one centimetre of the screw's length is called the PITCH and is measured in threads per centimetre. Figure 55 shows a screw with a pitch of 4 threads per centimetre. If the nut is held still and the screw is turned round once, it will travel forward 0.25 centimetre: 4 turns will be needed to make the screw travel forward 1 centimetre.

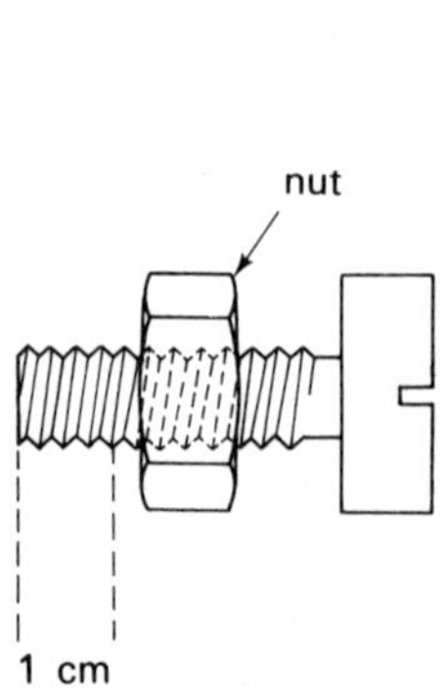

Figure 55

This is the principle of the screw jack, a simple variety of which is shown in Figure 56. A handle at the top of the screw is used to turn the screw round. The 'nut' to which a peg is fixed, which fits into a slot in the car, moves up or down depending upon which way the screw is turned. If the handle is 30 cm long, the effort applied at its end will move around a circle of 60 cm diameter for every one turn, a distance of about 188 cm. Whilst this is happening, the 'nut' moves only 0.25 cm. So the velocity ratio of the machine is very big:

$$\frac{\text{distance moved by effort}}{\text{distance moved by load}} = \frac{188 \text{ cm}}{0.25 \text{ cm}} = 752$$

This gives a very big mechanical advantage and explains why it is possible for a car to be 'jacked up' by even a child. Because of large frictional forces, the EFFICIENCY of a screw jack is very low. This has a practical advantage. If the machine has an efficiency of more than fifty per cent it will 'overhaul' (or run backwards) when the effort is released unless a special device is fitted to hold it in position.

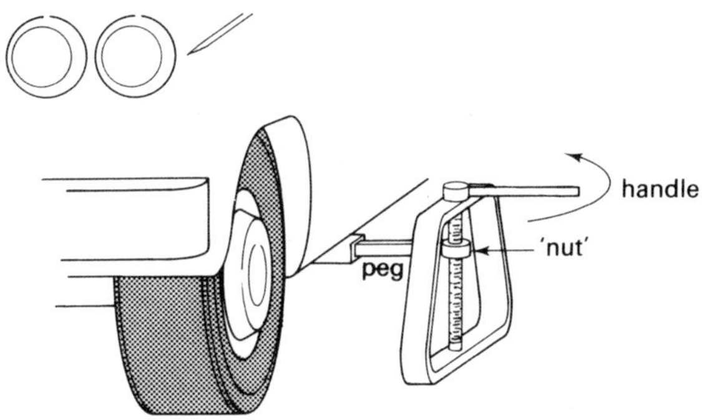

Figure 56

Self-test

ENERGY FORCE MACHINE LOAD EFFORT MECHANICAL ADVANTAGE EFFICIENCY VELOCITY RATIO MOMENTS LEVER SMALL LARGE FRICTION

Using the words in the list above, copy out and complete the sentences A to H:

A A machine does not save ______ but it enables a small ______ to be used to raise a big load.

B A ______ is a simple machine which depends upon the principle of ______.

C A machine can never have an ______ of 100 per cent: some of the energy transferred to the machine is used against the forces of

D In a machine, a small effort generally moves a ______ distance compared with the ______ distance which the load moves.

E If a load of 60 N is raised by an effort of 12 N, the ______ of the machine is 5.

F A screw jack is an example of a ______ with a very big ______.

G The velocity ratio of a pulley system is high if it uses a ______ number of pulleys.

H An inclined plane is a simple machine which gives a mechanical advantage because the ______ acting along the plane is less than that needed to raise the ______ directly upwards.

Questions

1 Draw a diagram which shows the moments of forces in a second-class lever. Give an example of a second-class lever as found in an everyday article.

2 Pulling the effort rope of a pulley system 4 m with a force of 20 N enables a load of 60 N to rise 1 m. Work out the mechanical advantage and the velocity ratio of the system.

3 Draw a sketch of a capstan, the device used by sailors on old ships to raise the anchor. Explain why a capstan is a simple machine.

4 Draw a diagram of a *hydraulic* jack in a simple form (see chapter 4). Explain how it provides a mechanical advantage and why it is more *efficient* than a screw jack.

5 Draw a diagram to show how the effort applied by the feet on a bicycle drives the machine forward and explain why the velocity ratio of the machine is less than 1.

9 Heat

It is easy to tell the difference between hot and cold: our sense of touch is able to feel differences of TEMPERATURE, which is the measurement of how hot or how cold anything may be. A thermometer is an instrument which measures 'hotness' and 'coldness' in degrees Celsius (°C). A normal laboratory thermometer is generally marked from −10°C to 110°C, covering the range of temperatures most likely to be met. Two points on this range are called 'fixed' points and are used for calibrating the thermometer. These points are:

0°C the temperature at which pure ice melts
100°C the temperature at which pure water boils (and turns into steam).

To be very accurate, it should be said that these temperatures are those recorded at an atmospheric pressure of 760 mm of mercury. Variations in pressure affect boiling points rather more than freezing points.

Figure 57 shows a diagram of a mercury-in-glass thermometer. An increase in temperature causes the mercury to EXPAND along the narrow bore of the tube. A rise in temperature causes all materials to expand, but different materials expand by different amounts for the same temperature rise.

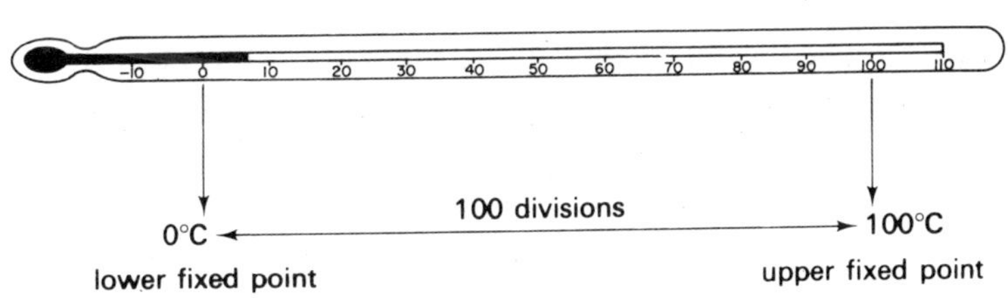

Figure 57

Mercury is a very expensive material but it has great advantages for use in a thermometer:

(a) it is a metal and heat travels through it easily
(b) it expands evenly over a wide range of temperatures
(c) it is easily seen in the narrow tube
(d) it does not 'stick' to the walls of the tube
(e) it freezes at −40°C and boils at 360°C – a good range for most purposes.

Alcohol, with a little dye to make it visible, is sometimes used in thermometers. It is much cheaper than mercury but it boils at 78°C which is lower than the boiling point of water and makes it unsuitable for laboratory use. It freezes at −114°C which makes it better than mercury for very cold temperature measurement.

The clinical thermometer (Figure 58) is a special type of mercury-in-glass thermometer for measuring the body temperature which, in normal health, is about 37°C but rises with feverish illness. It is really a MAXIMUM thermometer, one which records the highest temperature reached and leaves a mark at that temperature. The clinical thermometer is normally placed beneath a patient's tongue for a certain time and is then removed for reading. Since body temperature is higher than room temperature, an ordinary thermometer would show a fall in reading as soon as it was removed from the patient's mouth. In the clinical thermometer, the constriction (narrowing) causes the mercury thread to break as soon as cooling takes place, leaving the upper part of the thread in the tube to show the patient's temperature. Before it is used again, the thermometer is shaken to make the mercury thread join up again.

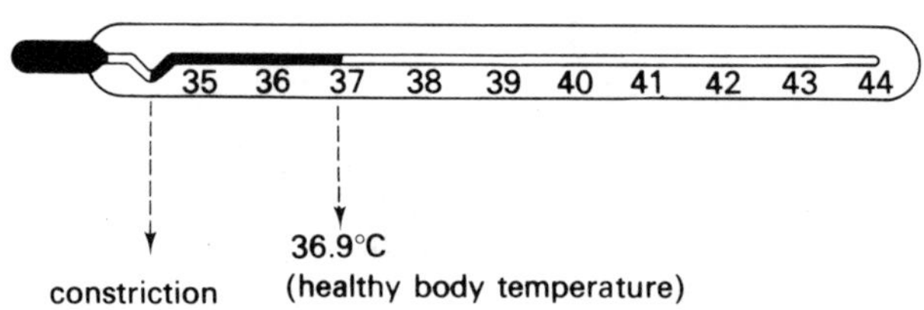

Figure 58

Expansion due to heat

The thermometers described depend upon the fact that mercury and alcohol expand when there is a rise in temperature. *All* materials, solid, liquid and gas, expand with a rise in temperature as will be shown by the following experiments.

Figure 59 illustrates a laboratory experiment to show the expansion in length of a metal rod. The rod, about 50 cm long, is

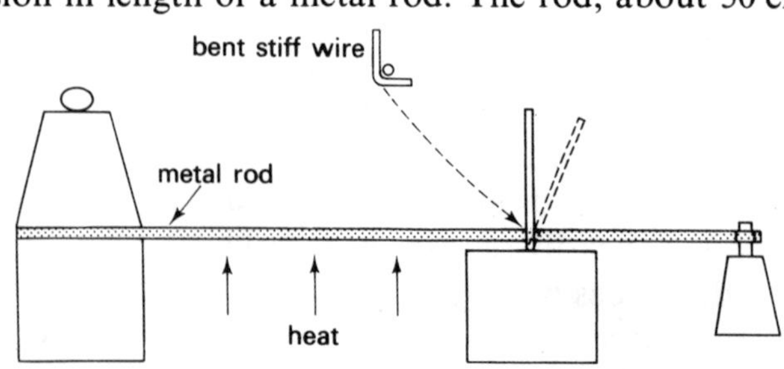

Figure 59

firmly fixed at one end and has, at the other end, a piece of stout wire bent at right angles underneath. Any slight movement of the rod will cause the wire to roll and the movement to be seen. This happens when the rod is heated, showing expansion.

Figure 60 is a bar which fits tightly into a gauge when cold. The bar will also fit the hole when cold. When heated it is found that the bar will not fit the gauge or the hole, showing that it has expanded both in length and in thickness.

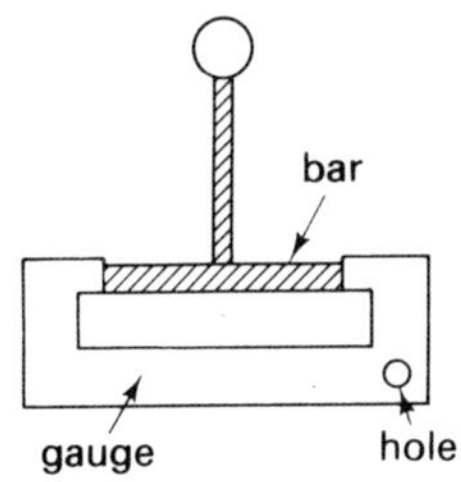

Figure 60

Figure 61 shows a flask completely full of coloured water and fitted with a narrow tube. Heat applied to the flask causes the water to rise quite quickly up the tube. Sometimes there is a slight drop in the water level before it begins to rise. This is due to slight expansion of the glass before the heat reaches the water.

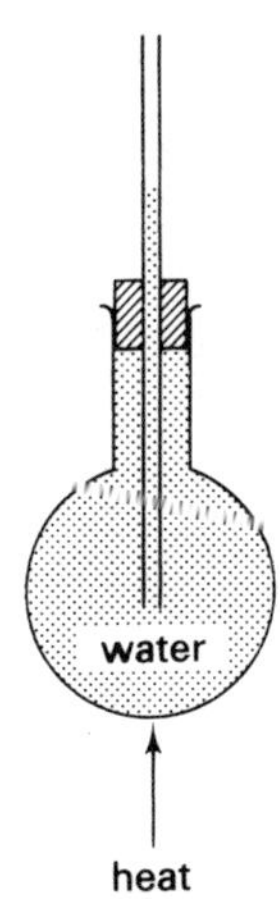

Figure 61

Figure 62 shows a way of investigating the expansion of a gas, air. A very slight rise in temperature, even that obtained by wrapping one's hands around the flask, causes the water in the 'U'-tube to rise up the right-hand limb. When the heat is removed, the gas soon contracts and the levels in the 'U'-tube return to their original position.

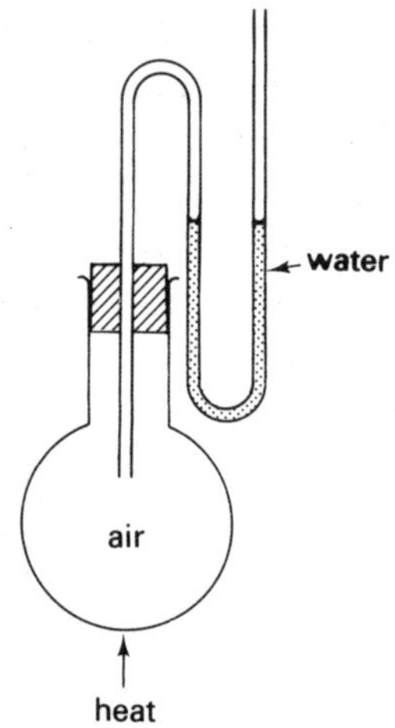

Figure 62

For the same rise in temperature:

solids expand least
liquids expand more
gases expand most.

When there is a fall in temperature, the materials CONTRACT in the same way.

Metals such as copper, iron, brass, aluminium, etc. all expand when heated but do not expand equally. A bimetallic strip (Figure 63) is made by firmly welding together two different metal strips, for example copper and iron, of the same dimensions. When the temperature increases, copper expands more than iron and the strip bends with the copper part on the *outside* of the curve.

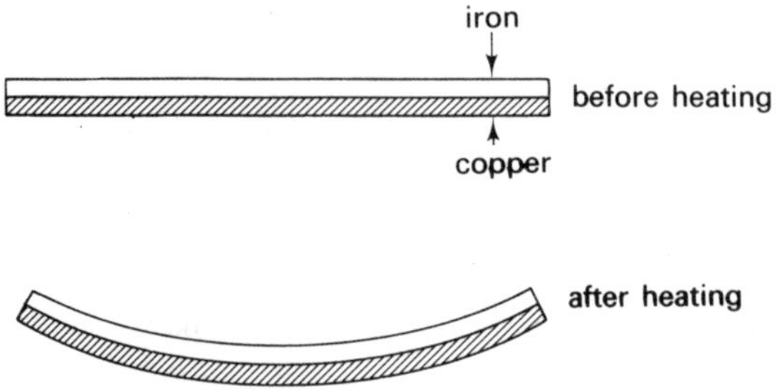

Figure 63

The forces exerted when materials expand and contract due to changes in temperature are enormous. They have to be taken into consideration in the design of buildings, bridges, railway lines and all types of machinery. Concrete roads expand in the heat of summer and gaps filled with pitch have to made between each 'slab' of concrete to allow the expansion to take place. Without these, the road would buckle upwards under the huge stress. The forces of expansion and contraction can be put to use in various ways. The steel 'tyres' fitted to the wheels of railway coaches are heated strongly to make them expand, fitted on to the wheel and then cooled. The contraction makes the tyre grip the wheel with great force and no further fixing is needed.

THERMOSTATS are used to control and regulate temperatures. They are used in central heating systems, electric blankets, ovens, tropical fish tanks, electric irons and so on. A gas oven thermostat is shown in Figure 64. Invar is the name of a metal alloy which expands very little for quite considerable rises in temperature. A rod of this metal is fixed to one end of a brass tube. The other end of the invar rod is fixed to a mushroom-shaped valve which, as it moves, controls the flow of gas to the oven burners. The brass tube inside the oven expands as it becomes hot and pulls the valve to cut down the gas supply. Should the oven temperature fall too much, the brass contracts and the valve lets more gas through. In this way the temperature of the oven can be kept at any required temperature, adjustment of the valve being made by a regulating knob.

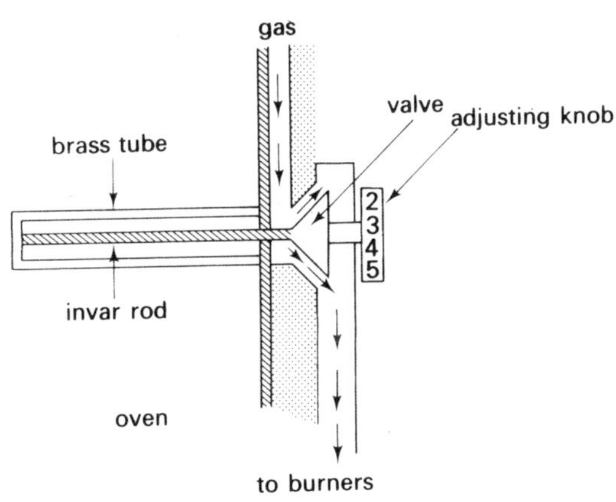

Figure 64

The bimetallic strip is ideally suited for use in thermostats. Figure 65 shows a simple example of one which could be used for controlling the temperature of an electric iron. The bimetallic strip forms a heat-operated switch. When the temperature rises, the bimetallic strip bends upwards, breaking the contacts and switching off the current to the heating element. Cooling down causes the strip to return to its original position and the current to be switched on again.

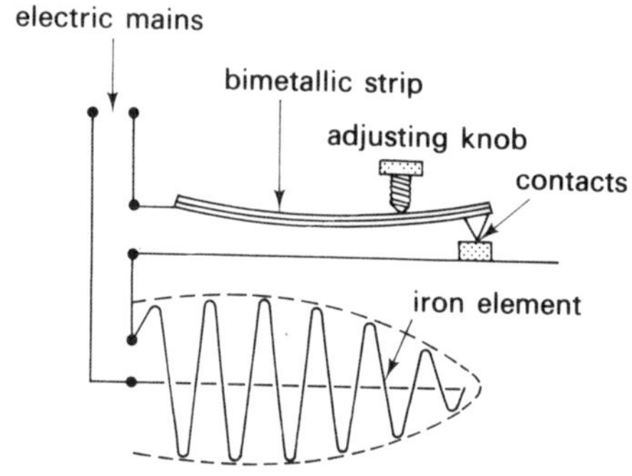

Figure 65

How heat travels

Heat always travels in one direction, from the higher temperature to the lower. Figure 66(a) shows a flask containing hot water. The flask is placed on a bench, a thermometer inserted and the temperature taken at one-minute intervals for fifteen minutes. The temperatures are plotted against time on a graph. The results are shown in Figure 66(b). At first, heat escapes quite rapidly but then, as the temperature of the water gets closer to the temperature of the room, the rate of cooling decreases.

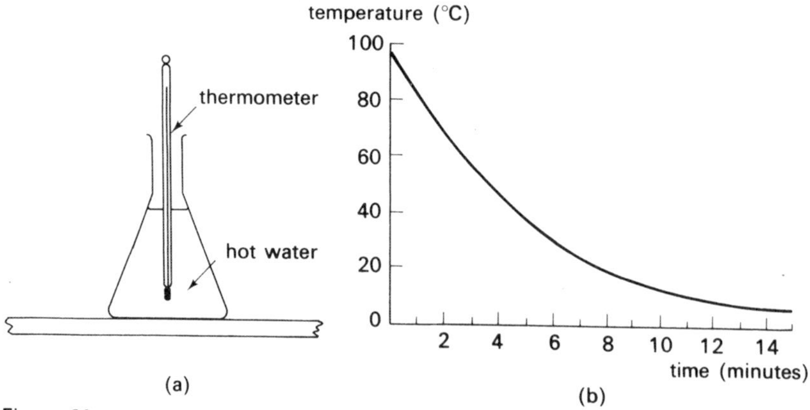

Figure 66

The cooling curve of the water shows that heat has travelled from the hot water to the air in the room but it does not show *how* heat travels.

If a metal rod is held between the fingers and the end placed in a flame, it is not very long before the rod has to be put down because it has become too hot to hold. If the test is carried out with a glass rod of exactly the same length, it will be a very long time before the rod needs to be put down. In fact, the end of the glass rod will melt long before this is necessary. It is also obvious that a match does not have to be dropped until the flame has burnt right along the wood. Heat is

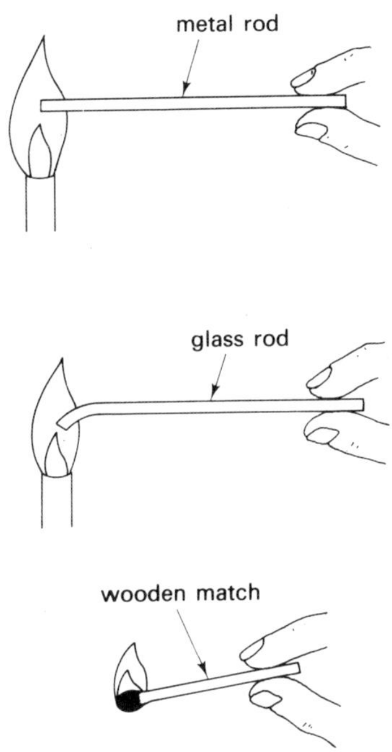

Figure 67

CONDUCTED more quickly along the metal rod in comparison to the glass and the wood. (See Figure 67.)

All metals are good heat conductors but all metals do not conduct heat at the same rate. How quickly they conduct heat can be investigated by experiment. An example of such an experiment is shown in Figure 68. Rods of different metals of the same length and thickness are bent into identical L-shapes and a pellet of soft wax is placed in the same position on each rod. The shorter length is placed into water, kept at its boiling point, and the time taken for heat to travel along the rod and melt the wax pellet, which falls off, is noted for each metal. Copper is a very good conductor of heat, better than most other common metals.

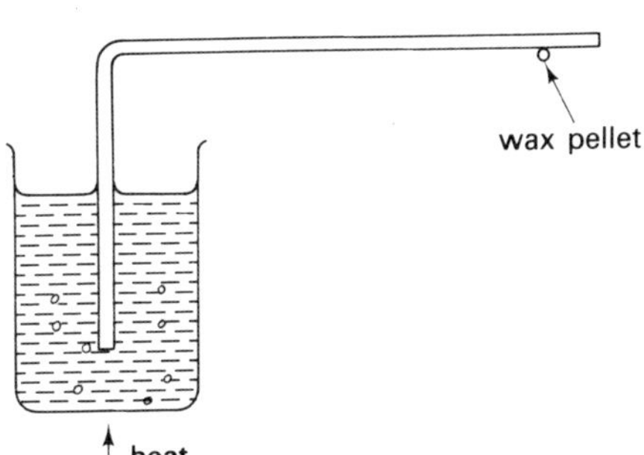

Figure 68

Non-metals are poor conductors of heat and so things like paper, glass fibre and glass wool, expanded polystyrene, sacking, vermiculite (a mineral material in very light pieces looking like a breakfast cereal), wood shavings and so on, are very useful materials for preventing heat from escaping to the cold atmosphere. This process, known as heat insulation, is very important in houses where every attempt has to be made to reduce the amount of valuable fuel burned to warm the house.

Air itself is a poor heat conductor and materials containing a lot of air are amongst the best of insulating materials. This is why double-glazing is used in homes. It is the air between the two sheets of glass that limits the loss of heat through the window. Figure 69 shows various examples of insulation used in everyday life.

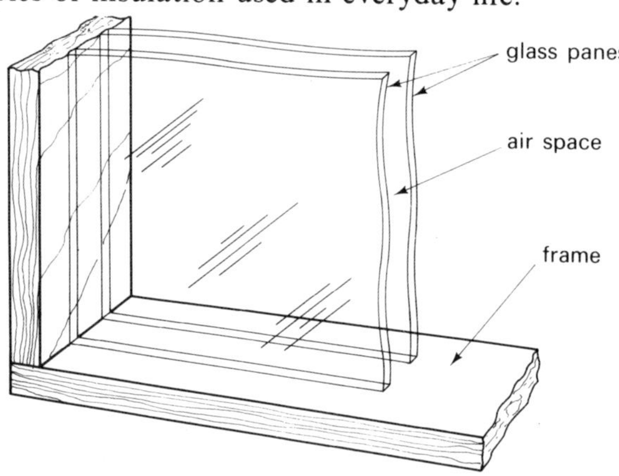

Double glazing

Insulated container for picnic food

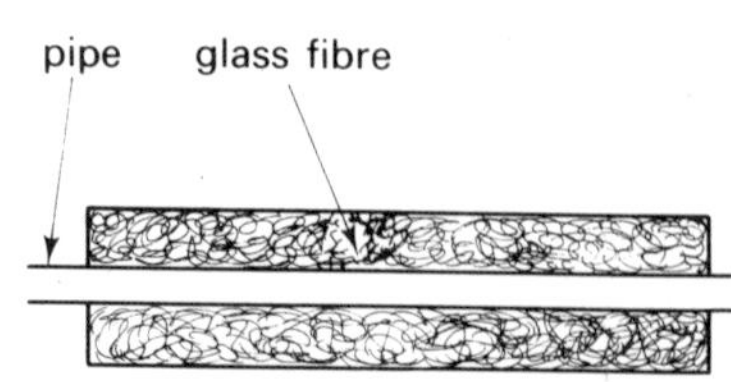

Lagging of water pipes to prevent freezing

Figure 69

Since air is a poor heat conductor, it would take a long time for heat to pass through several centimetres of air if it travelled only by CONDUCTION. If an electric fire is switched on, the heat from its element is felt quite a long way away particularly when the element becomes red hot. Also, heat from the sun reaches us across 93 million miles of space where there is no air until about 200 miles from the surface of the Earth. Heat travels through air and through space by RADIATION in much the same way as light. Like light, radiant heat can be reflected and most electric fires have curved reflectors behind the element to 'beam' the heat forward. An object placed in front of a source of radiant heat can cause a heat shadow as when a cloud passes before the sun and an immediate cooling effect is felt.

Different surfaces radiate heat at different rates. Figure 70 shows three identical glass flasks holding identical volumes of water at the same temperature, with a thermometer placed in each flask. Flask A is coated matt black with soot; flask B is left clear and flask C has a polished silvered surface. All flasks are in contact with the same surface and will lose heat to it by conduction equally. At the end of say, five minutes, the temperature in each flask is read and it is found to be highest in the silvered flask, lower in the clear flask and lowest in the matt black flask.

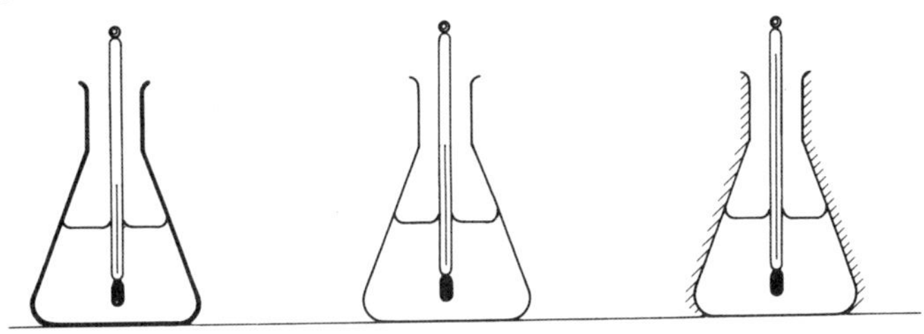

Figure 70

It would be wrong to jump to a conclusion on the basis of just one experiment, but, on the experience of many similar experiments with all precautions taken against error, it has been learned that a matt black surface is a better radiator of heat than a shiny, silvered one. Apart from the fact that a bright, silver teapot is attractive to look at, the silver surface helps to prevent loss of heat by radiation.

Figure 71 shows the same flasks, A, B and C, used in a different way. They contain the same volume of *cold* water and are placed at equal distances from a radiant heater worked by electricity. At the end of several minutes, the temperature of the water inside each flask is read and it shows that the water in the matt black flask is warmest, next warmest in the clear flask and the coolest is in the silvered flask. With other similar experiments carried out to check this, it becomes clear that dull black surfaces ABSORB radiant heat better than bright silver or white surfaces which reflect it. There is a *possible* connection between this fact and the reasons for wearing white clothing in the tropics and darker clothing in cooler climates.

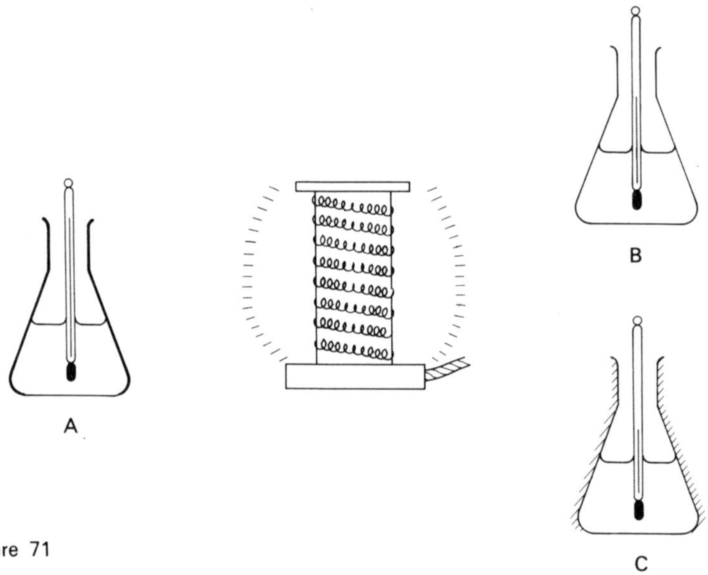

Figure 71

If one were to place one's hand in various positions at equal distances from a candle flame, it would be found that the greatest heat would be felt immediately *above* the candle flame. This is sometimes briefly explained by saying 'hot air rises' but it is worth considering *why* hot air rises and whether the same thing applies to liquids.

Figure 72 shows an experiment carried out with a few crystals of potassium permanganate placed at the bottom of a flask of cold water. Potassium permanganate turns water purple as it dissolves. When gentle heat is applied, threads of colour rise up through the water and then turn round and come down again.

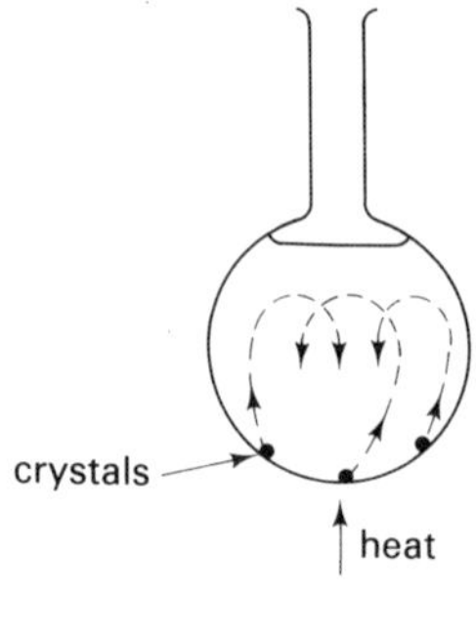

Figure 72

Figure 73 shows a glass-fronted box with two chimneys. Underneath one is placed a lighted candle. Smoke from a smouldering piece of string held at the top of the other chimney shows the currents of air: hot air rises up one chimney and cold air comes down the other to take its place. These currents are called CONVECTION currents and show another way in which heat can travel.

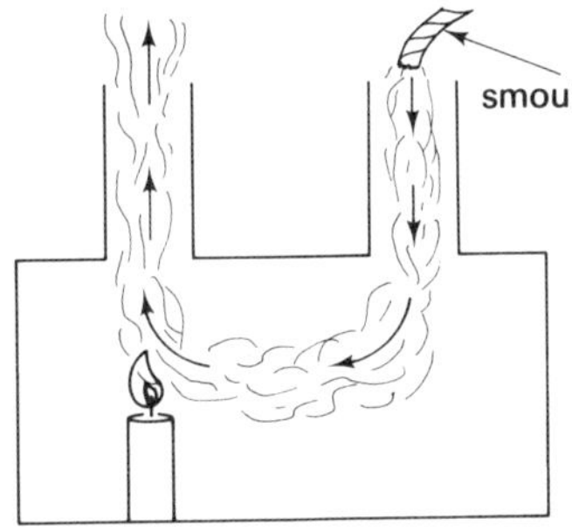

Figure 73

It has been shown that heat causes expansion of both liquids and gases. Expansion produces an increase in VOLUME. This means that the same MASS of the liquid or gas will take up more space: in other words it will become LESS DENSE and will rise as the denser, cooler liquid above it sinks down. As the less dense liquid rises, it cools, contracts, becomes denser and sinks down again.

Convection currents are responsible for draughts through the doors of houses as shown in Figure 74. To prevent the loss of heat by draughts, the cracks should be filled by rubber or plastic strip material.

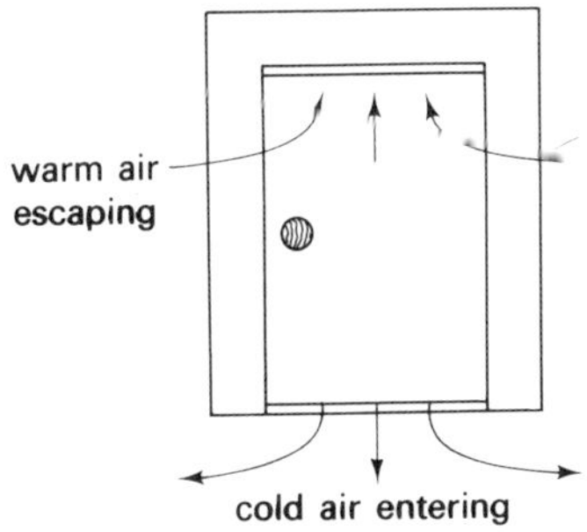

Figure 74

The vacuum flask

The vacuum flask (Figure 75) was invented by Sir James Dewar to keep liquid oxygen, a very cold substance, at a steady temperature. Vacuum flasks are now made by the million to keep liquids hot or cold for many hours. The transfer of heat by conduction is limited by the vacuum between the glass walls. Radiation is kept as low as possible by silvering the glass. If the flask is kept corked, heat loss by convection is limited.

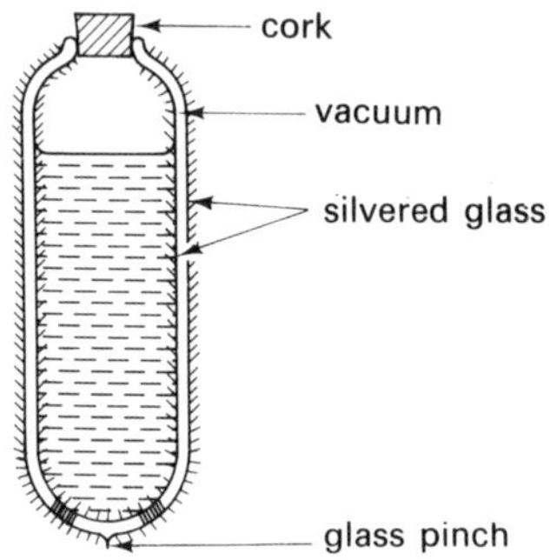

Figure 75

The quantity measurement of heat energy

Figure 76 shows two identical containers covered with an insulating material to reduce heat loss. A contains 0.5 kg of water at 10°C and B contains 1 kg of water at the same temperature. Each container has in it an electrical immersion heater of 50 W. (Each will transfer 50 joule of energy each second. See chapter 7.) Both are switched on for the same time and the rise in temperature noted. At the end of a period of time it is found that the temperature in A has risen by 4°C and the temperature in B has risen by 2°C. Each vessel has received the same amount of energy: the thermometer tells us only the difference in temperature – it cannot tell us that the water in each vessel has received the same amount of heat energy. To raise the temperature of the water in B by 4°C, it would be necessary to keep the immersion heater on for twice the time needed to heat the water in vessel A.

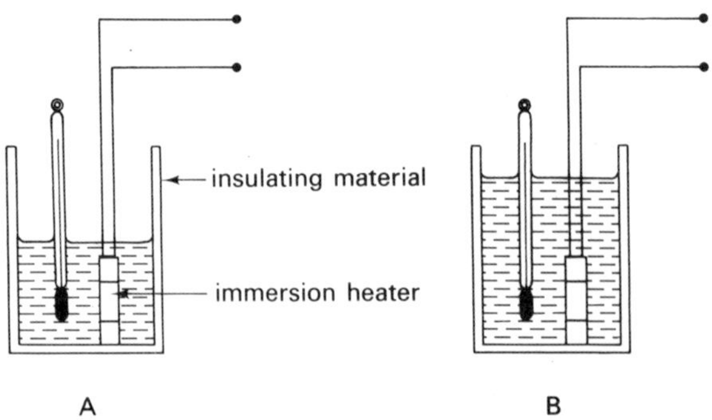

Figure 76

To raise the temperature of 1 kg of water by 1°C requires 4200 joule of heat energy; to do the same for 1 kg of alcohol requires 2400 joule. This measurement – the heat energy needed to raise the temperature of 1 kg of a material by 1°C – is called the SPECIFIC HEAT CAPACITY of that material and is written as joule per kg °C (J/kg °C).

In dietetics, the science of food, the number of CALORIES a particular quantity of a certain food provides when it is eaten is often stated. A calorie, in this case, is the equivalent of 4200 J. This is the amount of energy which that quantity of food provides for the human 'engine'.

Heat and changes of state

If solid lead is heated strongly, it melts and becomes liquid. It is still *lead* and has not been changed into another metal. When it cools down it becomes solid lead again. Water, if it is heated, eventually boils and becomes steam. Again, water and steam are exactly the same substance: when steam cools, it becomes water. Similarly if water is cooled it eventually becomes ice, a solid. When ice is warmed it becomes water again. There are many examples of these changes of STATE, brought about by changes in temperature.

If an immersion heater is placed into a beaker of water and switched on, the temperature of the water rises until it reaches boiling point and steam is produced. However long the immersion heater is left on, the temperature of the water does not rise any higher. Heat energy is steadily being supplied to the water but it is all used in changing water into steam *at the same temperature*, 100°C.

When a piece of ice at 0°C is gently heated, its temperature does not immediately rise. For a time all the heat energy is used in changing ice at 0°C into water at 0°C.

A change of state *without a rise in temperature* requires a certain amount of heat energy to bring that change about. This heat energy is called LATENT (hidden) heat.

To change 1 kg of water at 100°C into steam at 100°C takes 22.6×10^5 (2 260 000) joule. This is the latent heat of vaporisation of water.

To change 1 kg of ice at 0°C into water at 0°C takes 33.4×10^4 (334 000) joule. This is the latent heat of fusion of ice.

Evaporation

A puddle of water left by rain on an asphalt playground soon disappears. It has not soaked into the asphalt but it has evaporated or turned into vapour. Some liquids evaporate very quickly, like petrol or ether. They do not have to be *heated* for this change of state to take place but it happens more rapidly if there is a wide surface area of the liquid in contact with dry, warm and moving air. These are the best conditions for washing to dry on the washing line. At the surface, some liquid will always be going into the air in the form of vapour.

A change of state from liquid to vapour needs some heat energy and this is taken from the immediate surroundings. Wet clothing makes one feel cold because water from the clothing is turning into water vapour and taking the necessary heat from the body.

Figure 77 shows a simple experiment which is done with ether, a liquid which is VOLATILE (evaporates very easily). A little ether in a beaker is placed on a small pool of water on a block of wood. Air is bubbled rapidly through the ether, disturbing it, increasing the surface area, and making it evaporate quickly. The heat energy needed for the ether liquid to change into ether vapour is taken from the beaker and the water which soon freezes. The beaker can be lifted up with the block of wood frozen to it.

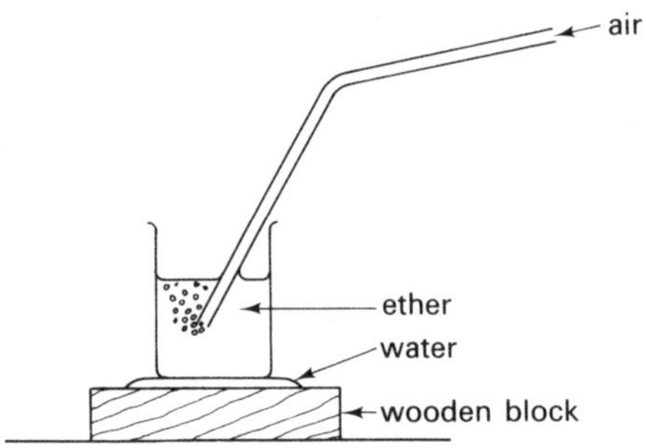

Figure 77

Figure 78 shows a simple device for keeping milk cool without a refrigerator. It depends again on the cooling due to evaporation.

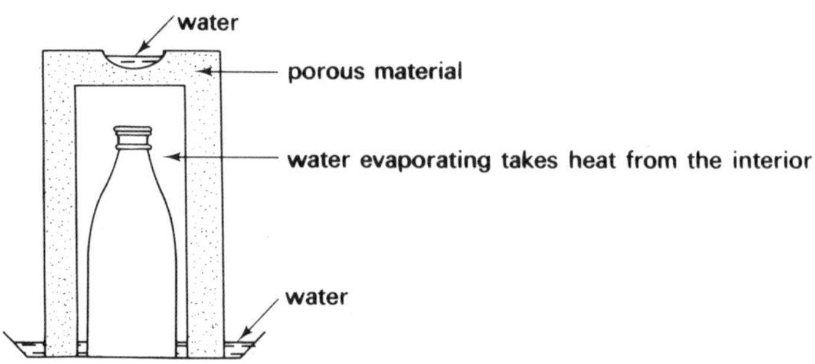

Figure 78

The effect of pressure on boiling point

Figure 79 shows a flask containing water and a thermometer. A side arm of the flask can be connected to a vacuum pump and some of the air removed. If the water is then heated, it will be found to boil at a temperature well below 100°C.

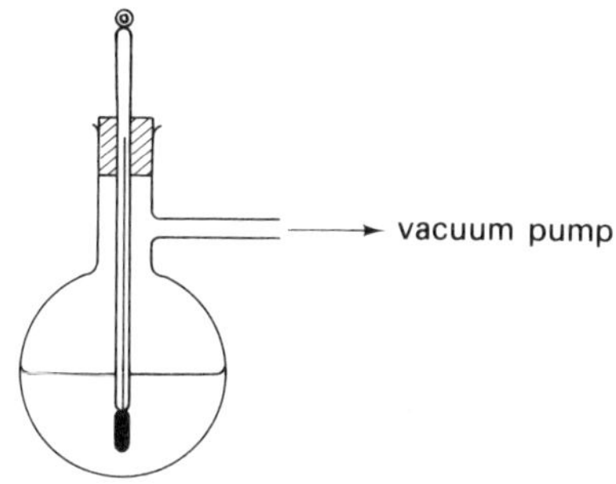

Figure 79

A reduction in air pressure makes the boiling point of all liquids lower. High up a mountain, where the air pressure is reduced, it is not possible to make a cup of hot tea by the usual method of boiling water in a kettle.

An increase in atmospheric pressure raises the boiling point of liquids. This is why a pressure cooker (Figure 80) enables such things as potatoes to be cooked in less time than it takes to cook them in a saucepan. In a pressure cooker, the potatoes are cooked in water which is considerably hotter than water boiling at 100°C. The increase in pressure in a pressure cooker is obtained by having a tight-fitting lid securely fixed to a strong saucepan. All pressure cookers are fitted with a safety valve which allows the steam to escape before it reaches a pressure which might be dangerous.

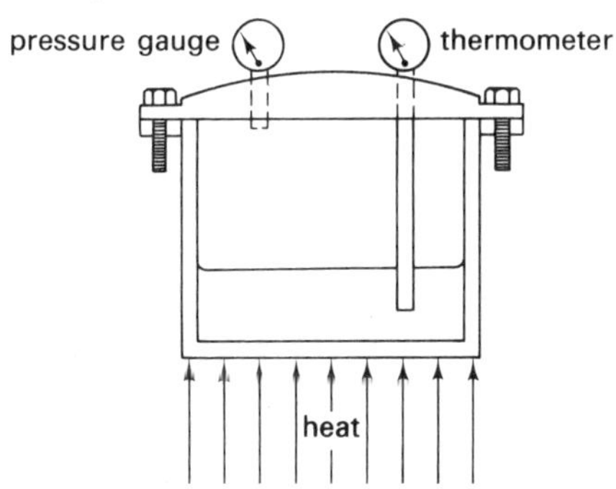

Figure 80 A pressure cooker

Why ice forms at the surface of water

When water turns into ice, the change of state of a liquid into a solid, there is expansion. Figure 81 shows what may happen when a bottle full of water freezes. 10 cm^3 of water at 0°C makes about 11 cm^3 of ice; so ice is less dense and floats. As its density is about 0.9 g/cm^3, it floats with 9/10ths of its volume under water. This makes it dangerous for a ship to sail close to an iceberg. Figure 82 shows how there is more of an iceberg under the water than is seen above it.

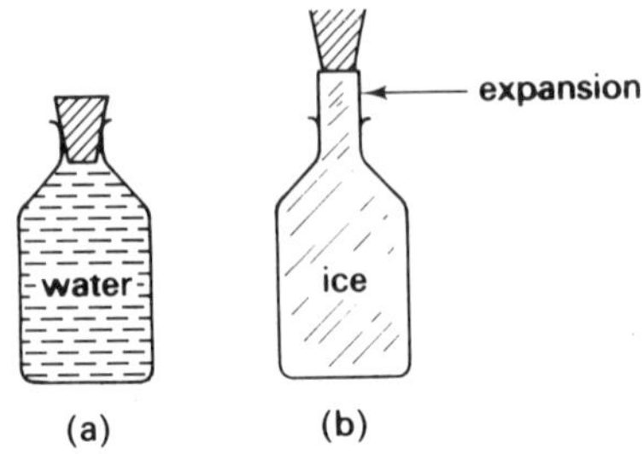

Figure 81

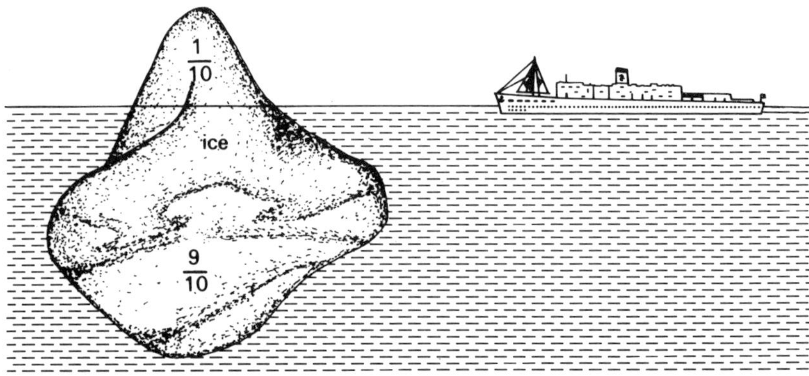

Figure 82

Because of this expansion, it can be seen why burst water pipes may occur in a house when there is severe cold and why the burst is only discovered when the ice has melted in the pipe!

When water is cooled it contracts. Its volume becomes less, so cold water is denser and sinks to the bottom. However, this contraction only takes place down to a temperature of 4°C. At temperatures lower than 4°C, water expands slightly, becomes less dense and rises to the surface. Here it becomes colder still in freezing conditions and eventually changes into ice. Figure 83 shows the temperatures in a pond which has frozen. The fact that water has its highest density at 4°C is of great importance to fish, whose survival depends upon the fact that water does not freeze from the bottom upwards.

Figure 83

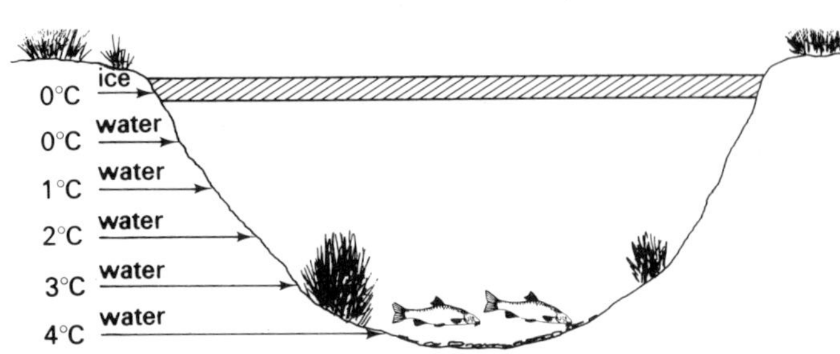

Self-test TEMPERATURE FREEZING BOILING CONDUCTION EXPANSION ALCOHOL VAPOUR MERCURY RADIATION CONVECTION DENSITY SPECIFIC HEAT CAPACITY EQUALLY STATE LATENT EVAPORATION PRESSURE INCREASE DECREASE

Using the words in the list above, copy out and complete the sentences A to J:

A A rise in temperature causes the ______ of solids, liquids and gases.
B A thermometer normally contains ______ but it may contain ______ if it is not needed to read temperatures above 75°C.
C All metals expand when heated but they do not expand ______ for the same rise in temperature.
D The ______ point of pure water is 100°C but an increase in pressure will cause this temperature to ______ .
E A certain amount of energy is needed to change the ______ of a substance without a change in temperature. This energy is called ______ heat.
F ______ is the change from liquid to vapour and it causes cooling due to the fact that ______ heat is required to make the change of ______ .
G Heat will travel slowly through solid materials by ______ .
H A matt black surface provides greater ______ of heat than a shiny, silvered surface.
I Heat cannot pass through a vacuum by ______ or by ______ .
J The energy needed to change the temperature of 1 kg of a substance by 1°C is called the ______ of that substance.

Questions

1 (i) Describe an experiment to show that metals expand when heated. (ii) Describe the action of a bimetallic strip when heated and explain how it shows that different metals do not expand equally when heated equally.
2 Give examples of the ways in which the loss of heat energy from a house by conduction and by convection may be reduced.
3 Explain carefully what is meant by latent heat. Why is more heat energy given out when steam cools from 100°C to 80°C than when the same mass of water cools from 90°C to 70°C?
4 Draw a diagram of a vacuum flask and explain how it reduces heat loss by conduction and by radiation.
5 Describe an experiment which could show that different amounts of heat energy are needed to raise the temperature of the same mass of water and of paraffin by the same amount.
6 Describe the way a pressure cooker is made and explain why it

enables potatoes to be cooked more quickly than in an open saucepan.

7 Explain carefully why the temperature of water in a hot water tank containing an immersion heater is usually quite cool at the bottom of the tank.

8 A quantity of wax is heated in a tube until it is all melted at 80°C. A thermometer is placed in the wax and the temperature is read every minute, continuing for ten minutes after the wax has turned solid. The temperature falls quite rapidly for five minutes and then remains at 56°C for four minutes when it begins to fall again. Sketch a graph to show the cooling of the wax and explain why the temperature remains the same for some time at 56°C.

9 A mercury-in-glass thermometer cannot be used to measure temperatures above about 350°C. Explain why this is and suggest a method by which high temperatures, in a furnace for example, might be measured.

10 Given samples of glass fibre, polystyrene foam chippings and sawdust, suggest an experiment to find out which of these would give the best results as a heat insulator to reduce loss of heat from a house.

10 Light

Light is a form of energy. This can be shown because, as with all other forms of energy, it can be changed from one sort to another. Figure 84 shows a solar cell with light falling on to it. Solar cells contain the metals selenium and gold which convert light energy into electrical energy. They are used on artificial satellites to provide electrical energy for radio equipment.

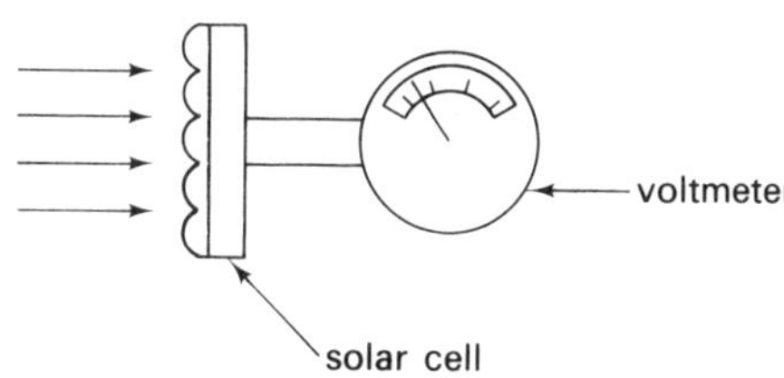

Figure 84

White light from the sun or from a filament lamp is not actually white as a single colour but is made up from light of several different colours, the colours which a rainbow shows. This was first discovered in an experiment by Sir Isaac Newton who passed a narrow pencil of light from a chink in a dark curtain through a triangular glass prism. The light comes out of the prism in a range of colours stretching from red to violet (see Figure 85). From red the colours go to orange, yellow, green, blue, indigo and then to violet forming what is called the SPECTRUM of white light.

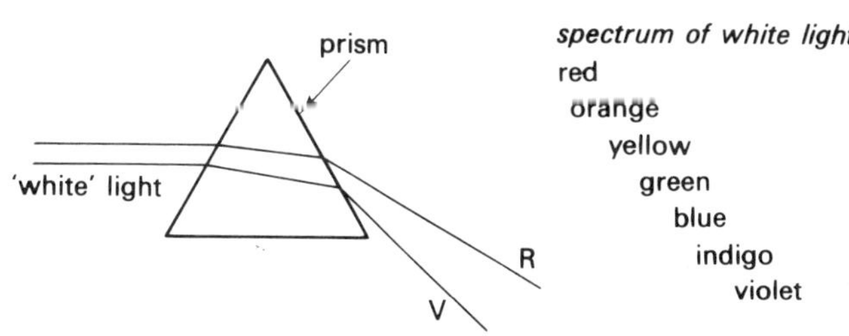

Figure 85

We are able to see because light coming from either the sun or a lamp of some sort is reflected and enters the eye. Light coming directly from, say, a star can also be seen and this is called INCIDENT LIGHT. Entering the eye the light is cast on to a sensitive lining on the back of the eyeball where light energy is changed into electrical energy and carried by a nerve to the brain. (See Figure 86.) Light coming from the lamp is reflected from the pot plant and enters the eye. *All* objects, except black ones, reflect light, not only highly polished surfaces as found in mirrors. The surfaces used in mirrors are very smooth and reflect light in a regular pattern, while the surface of other objects, being less smooth, reflect light but scatter it.

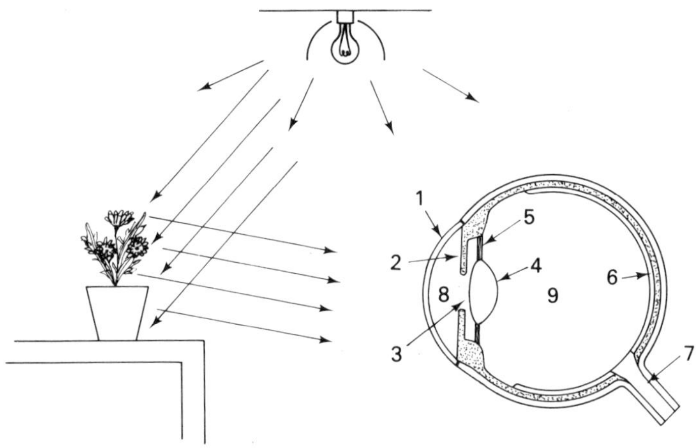

Figure 86

The parts of the eye in Figure 86 are numbered as follows:

(1) The CORNEA is the curved outer cover of the eye.

(2) The IRIS, the coloured part of the eye, is a sort of shutter. It controls the amount of light entering the eye, opening in dull conditions and closing in bright ones. The PUPIL (3) of the eye is really the hole left by the iris for light to enter.

(4) The LENS is shaped like a magnifying glass (a convex lens) but is of a material which can change shape when forces from the CILIARY MUSCLES (5) act on it. Acting with the cornea, it forms an image on the retina.

(6) The RETINA is the light-sensitive lining which changes light into electric 'signals' carried to the brain by the OPTIC NERVE (7).

(8) A watery liquid called the AQUEOUS HUMOUR fills the space between the cornea and the lens. Together with the cornea it acts as a convex lens.

(9) The eyeball is filled with a crystal clear jelly called the VITREOUS HUMOUR.

The eye sees colours because coloured objects contain PIGMENTS which reflect some of the colours of the white light falling on to them but ABSORB the others. Blue appears blue because it reflects the blue light but absorbs all the others. If pure yellow light fell on to a blue surface it would be absorbed, there would be no light to be reflected so the blue pigment would appear to be black. A black pigment absorbs all light and reflects none.

To see the pot plant in Figure 86 properly, a 'picture' of it must fall on to the retina of the eye. The 'picture' or IMAGE, as it is called, is formed by the lens. If a convex lens (Figure 87) is held close to a sheet of paper, movement of the lens, until just the right distance from the paper is found, will produce on the paper an image of a distant object. This can be done in a room using light coming from a window. (See Figure 88.) The image is smaller than the object and it is upside down (inverted). It is this image which falls on to the retina: it is the brain which turns it up the right way.

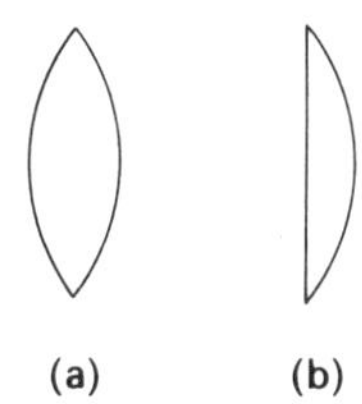

Figure 87 Convex lenses: (a) double convex, (b) plano convex

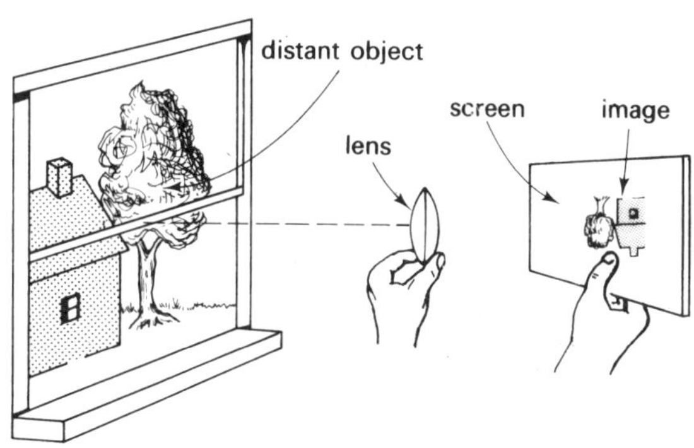

Figure 88

A camera obtains images in the same way; the lens makes an image fall on the film at the rear of the camera. A shutter across the lens allows just enough light to fall on the film to expose it properly: the shutter speed, the time the lens is allowed to be open, can be adjusted to allow a greater time when the light is dull. A camera which is needed to take photographs of close and distant subjects has to have an adjustment so that the image is sharp in both cases. This is done by moving the lens close to the film for distant subjects and further from it for subjects which are near. This can be shown by experiments using a lens, a brightly lit object, such as a lamp behind a screen with a cut-out arrow shape in it, and a screen. (Figure 89.)

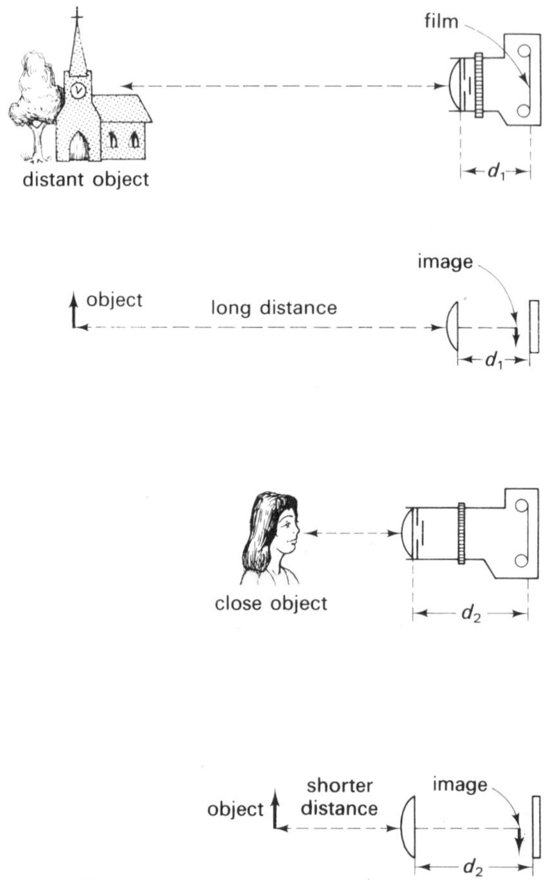

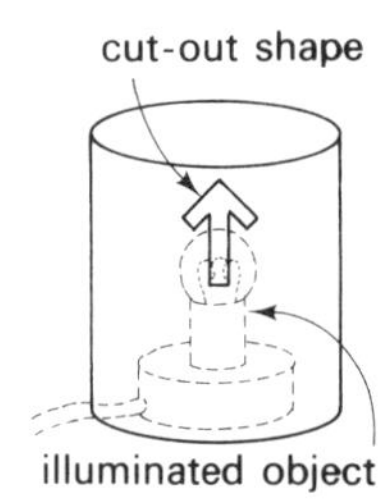

Figure 89

The human eye cannot focus on an object by moving the distance of the lens from the retina. Instead, the actual shape of the lens is changed to obtain a sharp image. This is done by the ciliary muscles.

Rays, beams or pencils of light (Figure 90) can be produced in the laboratory.

In schools these are generally obtained by using a 'ray box'.

Figure 90

The path which the rays take as they pass through lenses or are reflected from mirrors can then be traced, as shown in Figure 91 in the case of lenses and blocks of glass. The rays from the ray box, the INCIDENT rays, are shown coming from the left.

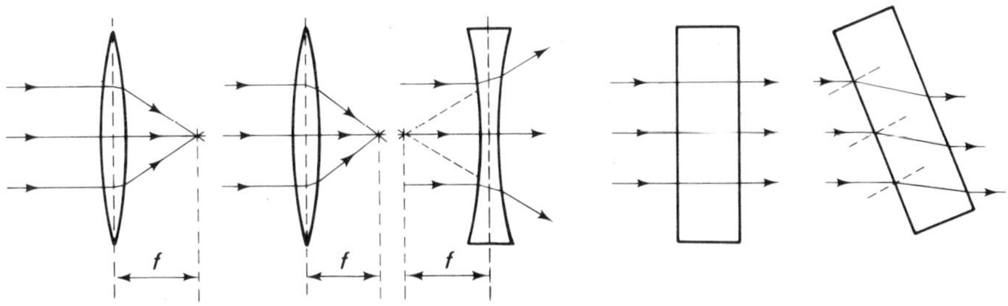

Figure 91 (a) (b) (c) (d) (e)

(a) A thin convex lens: the rays converge (come together) at a point a distance *f* from the centre of the lens.
(b) A thicker convex lens: the rays converge at a distance *f*, not so far from the centre of the lens as for the lens in (a).
(c) A concave lens: the rays diverge (go apart). The distance *f* can be measured only if dotted lines are drawn back from the diverging rays.
(d) A thick glass block: the rays meet the block at right angles and go on in a straight line.
(e) The same glass block as in (d) but the rays do not meet at right angles. They are 'bent' towards the dotted lines drawn at right angles to the surface of the block. When they come out they are bent again and take up a path parallel to the original one.

The distances marked *f* in Figure 91 are the focal lengths of the lenses: that is, the distance between the point where the rays of light meet and the centre of the lens. A thin lens has a longer FOCAL LENGTH than a thick one.

The paths of rays of light help to explain how images are formed. In Figure 92 the paths of two pairs of rays are traced from the top and bottom of an object. Rays a_1 and a_2 parallel to a line drawn at right angles through the centre of the lens change their paths to go through F, the FOCUS of the lens. Rays b_1 and b_2 going through the centre of the lens do not change direction. Where the two pairs of rays, a_1 and b_1, a_2 and b_2, come together, the top and bottom of the image are formed. Such an image, an inverted one and on the other side of the lens from the object, is called a REAL image. Real images can, like those made by a slide projector, be seen on a screen. The image seen in a looking-glass (a plane mirror) is the right way up, appears to be behind the mirror and cannot be put on to a screen.

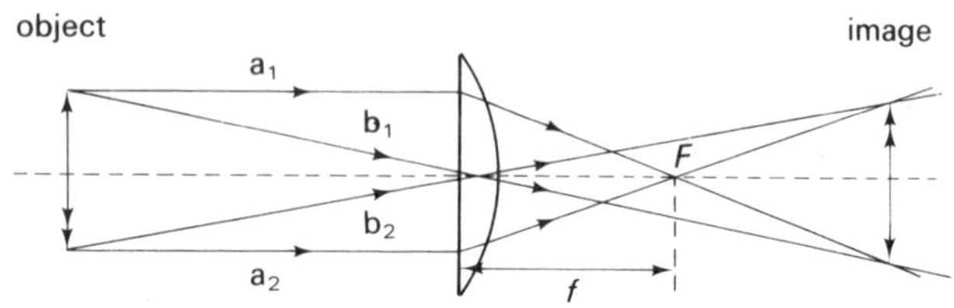

Figure 92

Lenses change the path of light rays because light travels more slowly through glass than through air. The speed of light is enormous, 3×10^8 m/s (186 000 m.p.s.). It is slightly less through denser materials like water or glass. When coming to a surface at right angles (Figure 91(d)), the light slows down but does not change direction. A ray doing this is called NORMAL. If the ray enters at an angle other than normal, it changes direction towards normal; when it comes out of the glass and into air, it speeds up again and changes direction away from normal.

The changing of direction of a ray of light when going from one material into another is called REFRACTION. Figure 93 shows light rays coming from a direction the eye believes to be straight from an object at the bottom of the tank. This is why the depth of water always appears to be shallower than it really is.

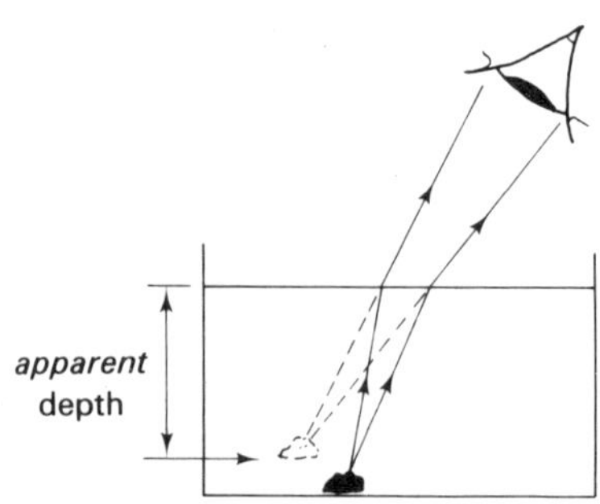

Figure 93

Using lenses

A convex lens can produce an enlarged, real image if the object is placed close to the focus but *not* between the lens and the focus. Figure 94 shows how, in a very simplified form, this can make a slide projector. The curved mirror behind the lamp concentrates light through the lens. The large, real image appears, as do all real images, upside down. The *slide* is put into the projector upside down to correct this.

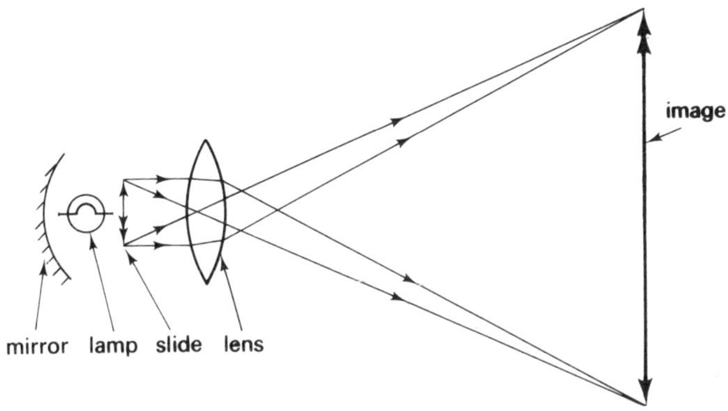

Figure 94 A slide projector

A magnifying glass

If an object is placed closer to a convex lens than its focal length, the lens produces a VIRTUAL image. This is on the same side of the lens as the object; it is upright and enlarged and cannot be shown on a screen. Rays traced from the object through the lens take the same paths as they do for a real image. The eye is deceived in believing that these rays from the lens travel in straight lines. (See Figure 95.)

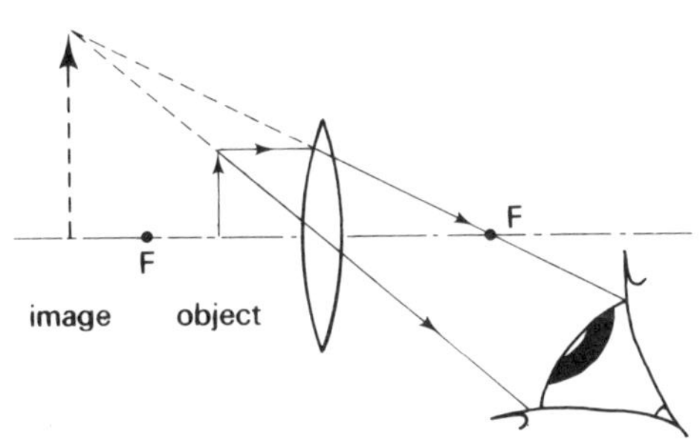

Figure 95 A magnifying glass

The astronomical telescope

A simple astronomical telescope (Figure 96) is made from two convex lenses: the objective lens (O) is a thin one with a long focal length and the eyepiece (E) has a short focal length. Light from a very distant object forms a REAL image (I_1). This does not fall on a screen but 'hangs in space' *just inside the focal length of the eyepiece*. The eyepiece then acts as a magnifying glass (Figure 95) and produces an enlarged virtual image. This image is upside down but it does not matter if the telescope is used only for stars and planets. A telescope for observing distant objects on the Earth's surface has another lens which makes the final image appear the right way up.

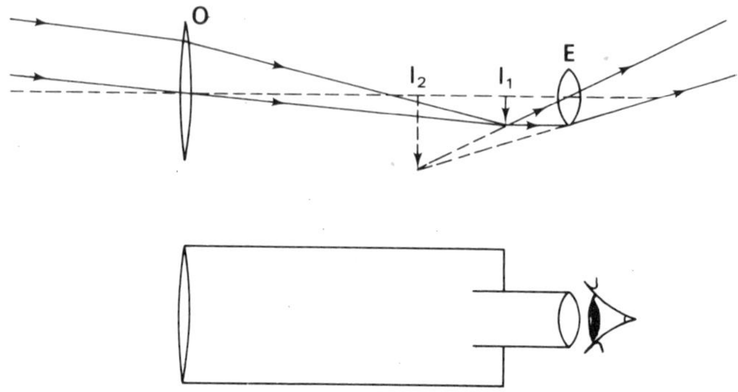

Figure 96 An astronomical telescope

Spectacles

Short and long sight (Figure 97) are caused because a defect of the eye does not allow the image to fall properly on the retina. To correct long sight, a convex lens is used to make the light converge: the rays come together in the right place. For short sight a concave lens makes the rays diverge for the same purpose.

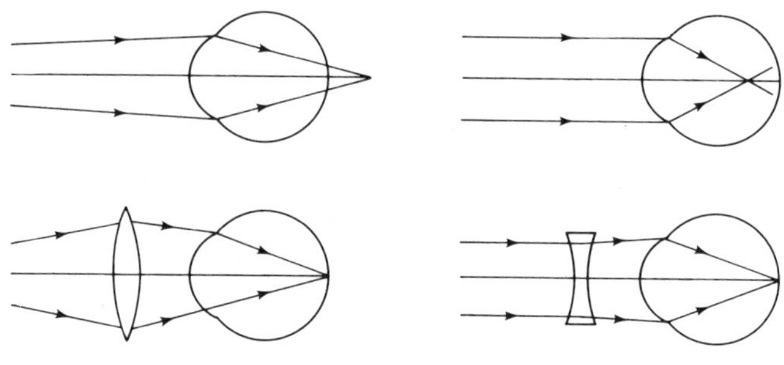

Figure 97 Spectacles

Reflection and mirrors

A very highly polished, smooth surface reflects the light falling on to it in a regular pattern. Ordinary glass, smooth water and polished sheets of metal, as well as mirrors, do this. Ordinary smooth surfaces, such as a sheet of paper, reflect light but not in a regular pattern. (See Figure 98.)

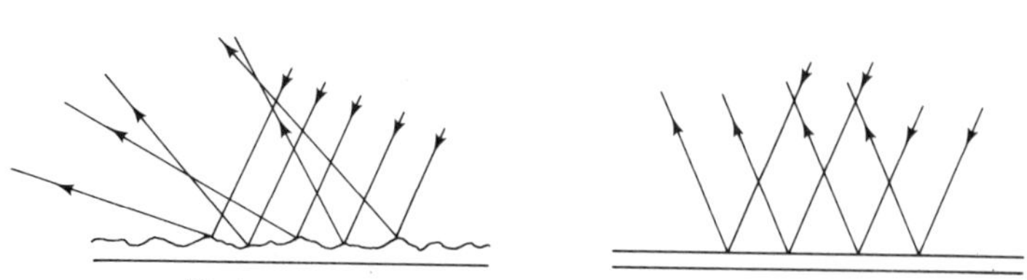

Figure 98

A single ray of light meeting a PLANE (flat) mirror is shown in Figure 99. The angle at which it meets the mirror, the angle of incidence (i), is equal to the angle of reflection (r). These angles are measured between the rays and the NORMAL, a line drawn at right angles to the mirror. In a mirror, the actual reflection is from a thin film of silver alloy on the back of the glass. Because of the reflection, the eye believes the reflected rays come from behind the mirror and the image seen appears to be the same distance behind the mirror as the object is in front of it. (See Figure 100.)

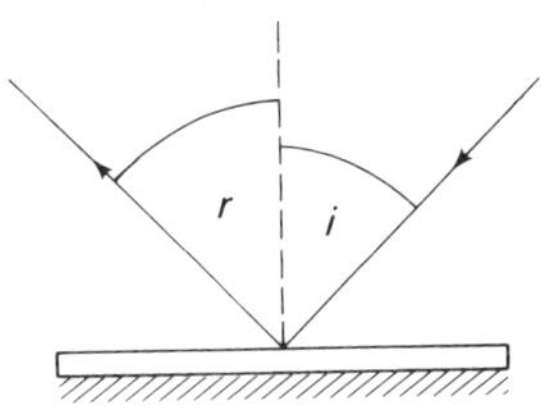

Figure 99

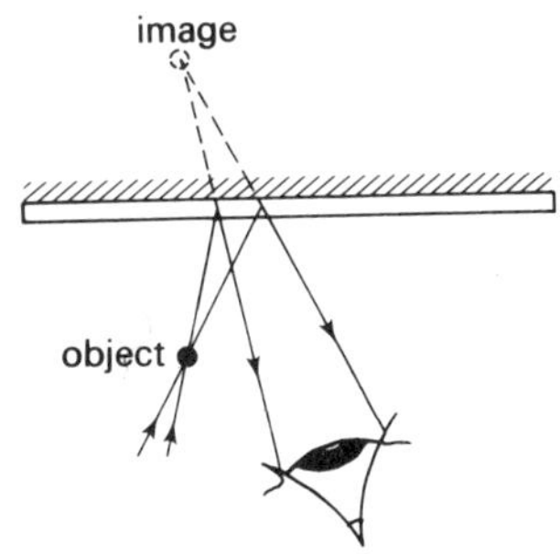

Figure 100

The image seen in a plane mirror is 'the wrong way round'; this is called LATERAL INVERSION (Figure 101). The image of your face in a mirror is not the same as is seen by someone looking directly at you.

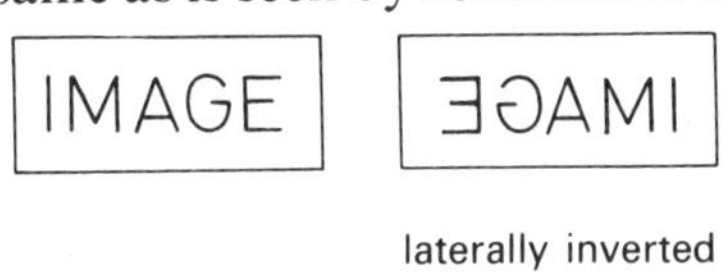

Figure 101

Curved mirrors are either concave or convex. The path taken by parallel rays when reflected is shown in Figure 102. A concave mirror makes the rays CONVERGE through the focus: the convex mirror make the rays diverge as though they came from a focus behind the mirror.

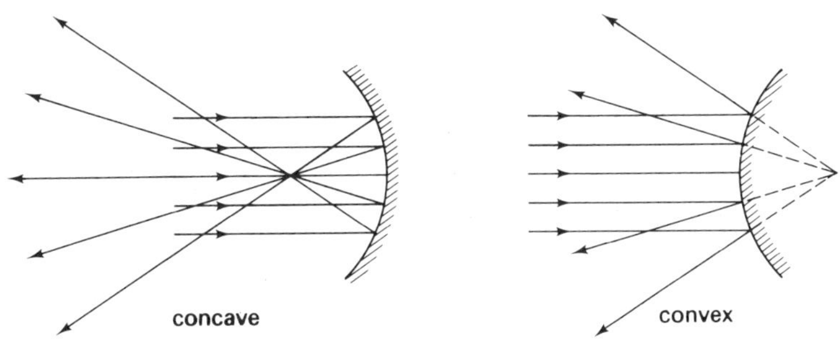

Figure 102

Big concave mirrors do not reflect *all* parallel rays through the focus. Rays towards the edge of the mirror are reflected short of the focus. For this reason, a lamp placed at the focus of a concave mirror would not produce the parallel beam of light needed for a flashlamp or a car headlamp (Figure 103(a)). A mirror in the shape of a parabola does give a parallel beam and is used as a reflector because of this (Figure 103(b)).

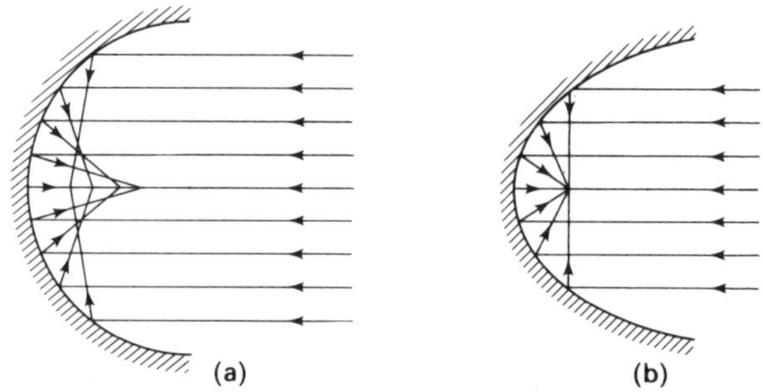

Figure 103

A concave mirror can produce a REAL image of an object at a distance and a VIRTUAL image of an object placed closer to it than its focus. A mirror which is very accurately made can be used in a reflecting telescope. The world's largest telescopes are of this type. The real image of the distant object produced by the mirror is enlarged by convex lenses in the *side* of the telescope. (See Figure 104.)

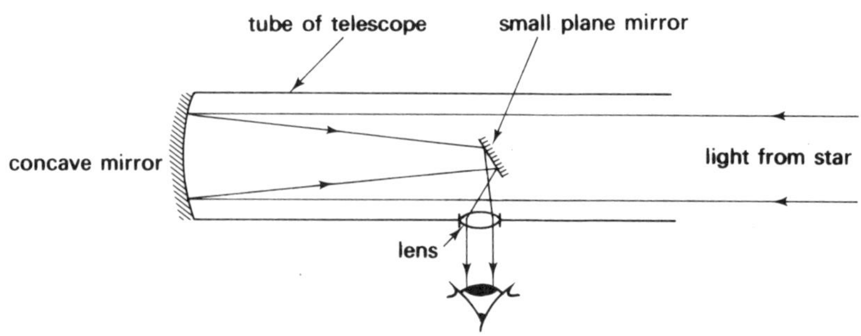

Figure 104 A reflecting telescope

A concave mirror producing a VIRTUAL image is used in a magnifying shaving mirror and a small concave mirror, fixed to a thin handle, is used by dentists to examine the back of teeth.

A convex mirror will only produce virtual images which are diminished (smaller). This type of mirror is often used for car wing mirrors because they give the driver a very wide view of the road and traffic behind him.

Reflection without mirrors

Refraction, as in the case of lenses, glass blocks and so on, is the bending of light rays due to a change in their speed. When a ray is travelling through glass and comes out into air, it bends *away from normal* as its speed increases. Figure 105 shows rays passed into a glass semicircle at different angles measured from the normal. In Figure 105(a) the ray is bent away from normal. In Figure 105(b), as the angle *i* increases, the ray is bent so much that it goes along the surface. In Figure 105(c), with a further increase in the angle *i*, the ray is reflected. The angle at which the ray does not come out of the glass, (Figure 105(b)), is called the CRITICAL ANGLE and, for glass, this is about 42°. If a ray 'inside' glass meets the surface at an incident angle of more than 42°, it will be reflected. Reflection by this means is better than from a mirror surface because less of the light is absorbed.

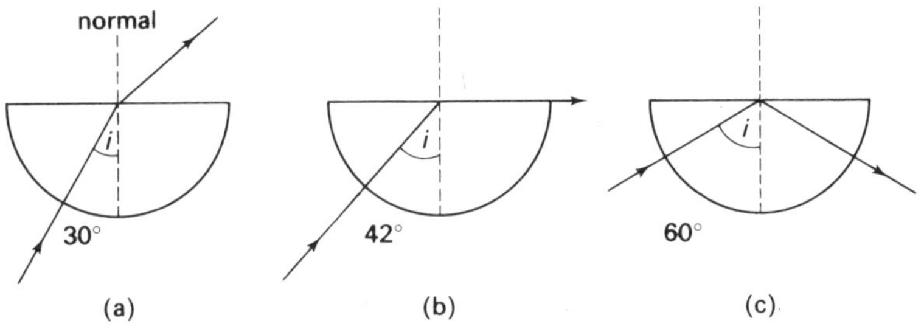

Figure 105

Figures 106–108 show how INTERNAL REFLECTION is used in a periscope, binoculars and a camera.

A simple periscope can be made with two plane mirrors (Figure 106(b)) but two right-angled triangular glass prisms are used in better quality instruments (Figure 106(a)).

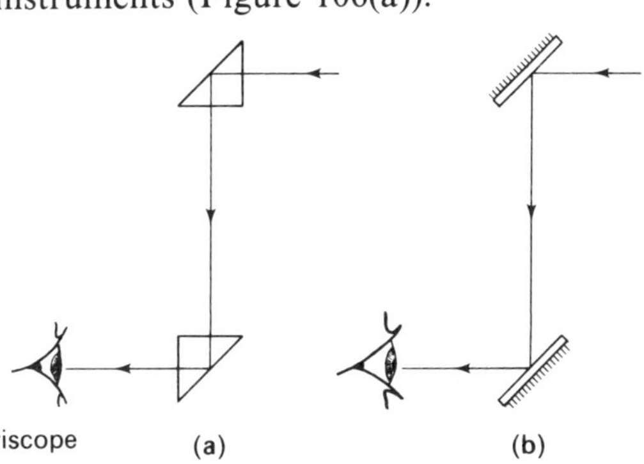

Figure 106 A periscope

Binoculars are a compact way of using two telescopes, one for each eye. Since telescopes are long and unwieldy, the use of two right-angled glass prisms, as shown in a simplified form (Figure 107), enables the distance between the eyepiece and the objective lenses to be packed into quite a short tube.

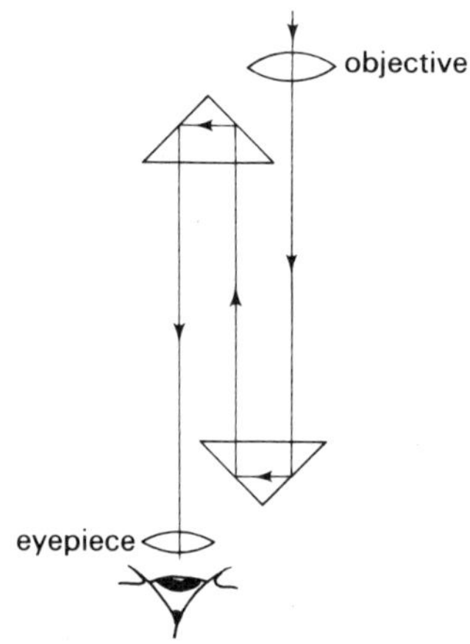

Figure 107 Prismatic binoculars

Figure 108 shows how the eye can see the photograph to be taken in a single lens reflex camera. When the image has been focused properly by movement of the lens, the mirror flicks out of the way to expose the film.

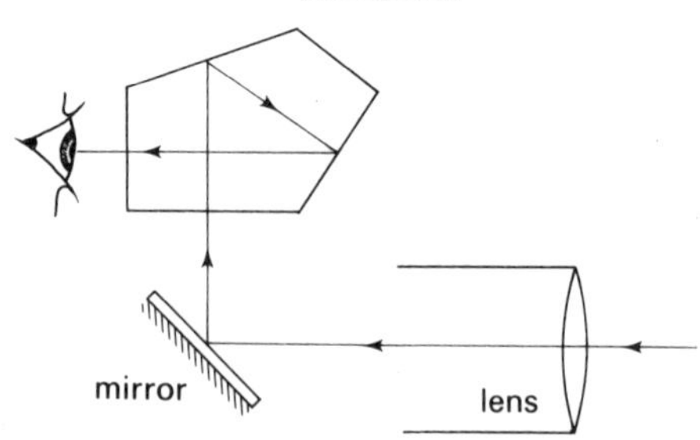

Figure 108 A single lens reflex camera

Self-test

REFLECTS ABSORBS CONVEX CONCAVE REFRACTION
SPEED SPECTRUM IMAGE REAL VIRTUAL
CONVERGE DIVERGE FOCAL LENGTH CRITICAL
INCIDENCE REFLECTION PARALLEL COLOUR

Using the words in the list above, copy out and complete the sentences A to J:

A The cornea and the lens of the eye produce an ______ on the retina.

B A pigment is a material which ______ certain colours and ______ others.

C The ______ of white light may be shown by the double ______ of a pencil of light as it passes through a prism.

D A ______ image is always upside down and it can, unlike a ______ image, be cast on to a screen.

E A convex lens held closer to an object than its ______ produces a magnified ______ image.

F Parallel rays of light passing through a convex lens are made to ______ .

G When a ray of light meets a plane mirror, the angle of ______ is equal to the angle of ______ .

H If a ray of light in glass meets the surface at an angle greater than the ______ angle, it is reflected.

I A ______ mirror can produce both real and virtual images.

J In a simple astronomical telescope the objective lens is a ______ lens with a long ______ .

Questions

1 Draw a labelled diagram of the human eye and explain (i) how it adjusts for close and distant objects and (ii) how the amount of light entering the eye is controlled.

2 Explain carefully why a red material appears different when looked at in daylight and under a blue lamp.

3 Explain how you would select, by simple experiment, which convex lens from a number of similar looking lenses, has the shortest focal length.

4 Draw a diagram of a simple *box* camera to show the position of the lens, the film, the shutter and aperture control. Explain why it cannot be used for photographing very close objects.

5 Draw diagrams to illustrate the defects of short and long sight and how they are corrected by spectacles. How would you find out by examining someone's spectacles whether they were for short or for long sight?

6 A convex lens has a focal length of 10 cm. By scale drawings find the position of an object (an arrow) placed (i) 20 cm and (ii) 5 cm from the centre of the lens.

7 A powerful flashlamp produces a parallel beam of light by the use of a reflector. Draw a diagram to show the shape of the reflector and the position of the lamp. Explain why the reflector is *not* shaped as part of a sphere.

8 When a thick plate of glass is placed on a printed page, the print appears to be slightly closer to the eye than before. Draw a diagram to explain this showing the paths of rays.

9 Compare the way (i) the human eye and (ii) a camera adjust to obtain sharp images of distant and close objects.

10 Draw diagrams to show how a prism of a certain shape can turn a ray of light through (i) 90° and (ii) 180°.

11 Electricity

Although an electrical current flowing through a wire cannot be seen, its effects can be observed. Electricity has three effects: (i) heating (ii) magnetic and (iii) chemical. Whenever electricity flows it has a heating and a magnetic effect, though both may be too weak to be noticed. The chemical effect happens only in special arrangements.

A battery, such as is found in a flashlamp, may be used to produce an electrical current. The word 'battery' really means a collection of similar units. A battery is made up from CELLS. Figure 109 shows how a cell is made. The electric current does not come from the cell.

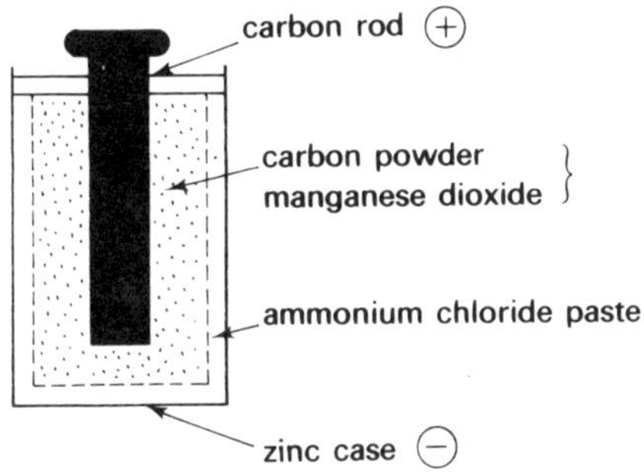

Figure 109 A dry cell

An electrical current is movement of ELECTRONS. The atoms of the materials, such as metals, which conduct electricity are made up of particles called protons and electrons. The electrons, which have what is called a negative electrical charge, can, if a 'pressure' is applied, be made to move from one atom to another through the material. The chemical energy in a dry cell provides that electrical pressure, which is measured in volt (V). One cell gives 1.5 volt. If a piece of wire is connected from the positive of the cell to the negative, the pressure of the cell makes the electrons in the wire move or 'flow'. To stop the flow of electricity, the wire circuit is broken by a switch. Figure 110 shows a flashlamp which contains two cells joined so the

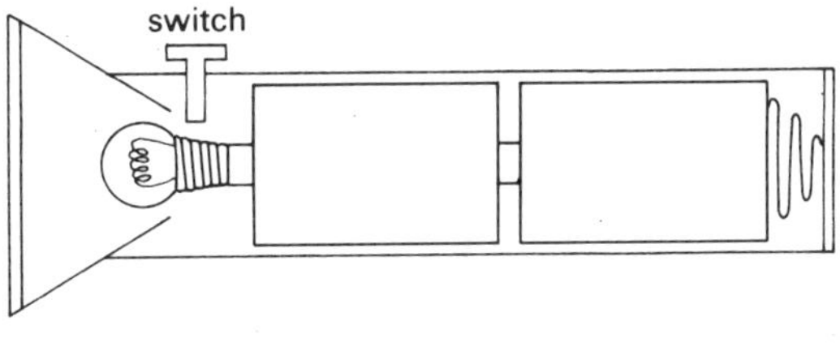

Figure 110 A flashlamp

positive of one is connected to the negative of the other. This gives an electrical pressure of $2 \times 1.5 = 3$ V. Higher voltage batteries are made by joining more cells in the same way; this is called SERIES connection.

Figure 111 shows the symbols used to illustrate the parts of an electrical circuit. Figure 111(a) is a circuit to show, simply, the heating effect of the current. The current which is taken to flow from + to – goes through the fine wire filament of the lamp and makes it white hot, providing light as a result.

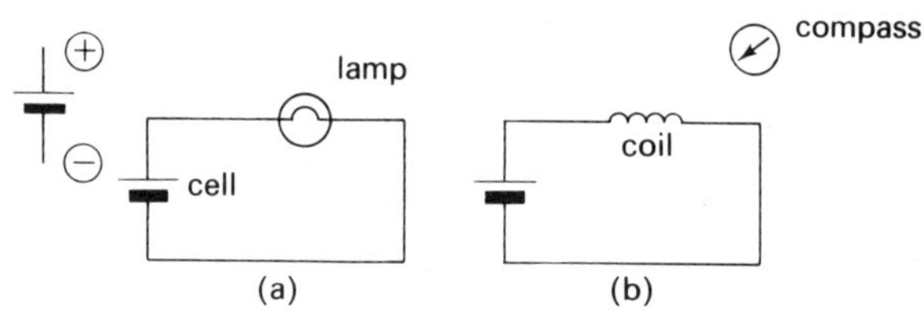

Figure 111

In Figure 111(b), the current goes through a coil of wire and a compass needle close to it is deflected as though the coil were like a bar magnet. If the direction of the current is changed by turning the cell around, the compass needle is deflected as though the magnet had been turned round. The magnetic force can be made stronger if INSULATED wire (with a plastic covering) is wrapped around a soft iron rod and more current made to flow by using more cells (Figure 112). When the current is switched off the iron ceases to be a magnet.

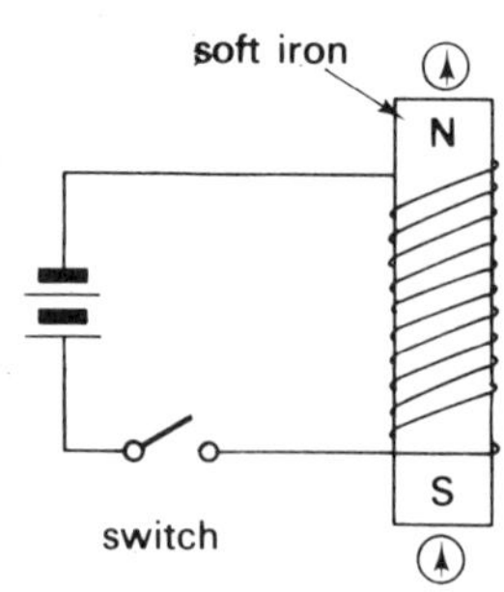

Figure 112

When an electric current flows through certain solutions or solids which have been made liquid by heat, it can split these materials into their elements. If a current is passed through water containing a little acid, water is split up into its elements, hydrogen and oxygen. The current is passed through the water by carbon ELECTRODES. Hydrogen collects at the negative electrode (called the cathode) and oxygen at the positive electrode, (the anode). This is a chemical effect of a current. It can be shown using simple apparatus (Figure 113).

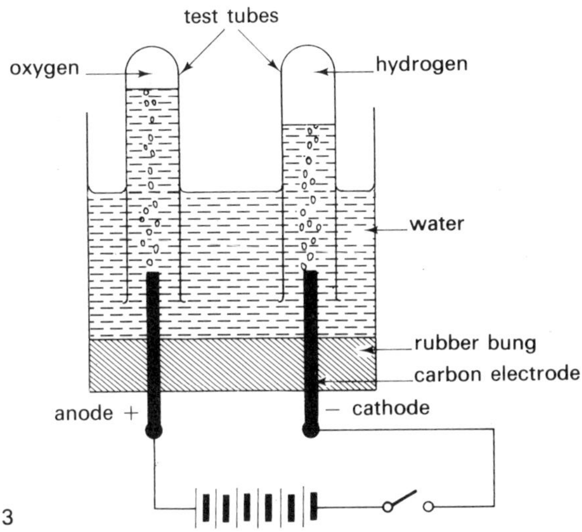

Figure 113

Simple lamp circuits

Figure 114 shows a number of circuits which can be made using suitable lamps (1.25 V, 0.2 A) and cells. If connected to one cell, the lamp lights normally. If connected to two cells it lights brightly and if it is connected to a supply of less than 1.5 V it will be dim.

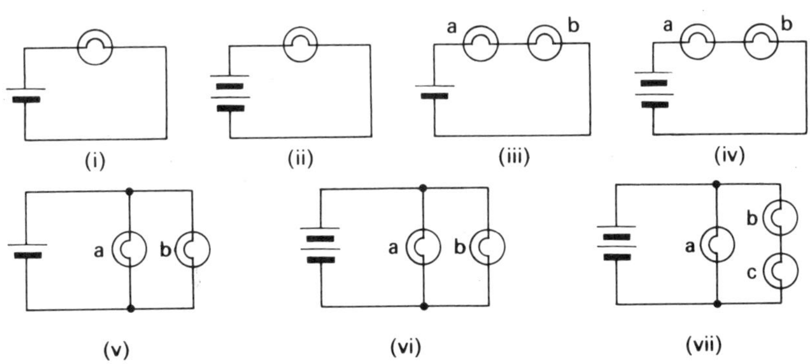

Figure 114

The results obtained with each of the circuits should be as follows.

(i) Lamp lights normally.

(ii) Lamp lights brightly.

(iii) Lamps (a) and (b) are dim. If either is removed, the other lamp goes out.

(iv) Lamps (a) and (b) light normally. If one is removed the other goes out.

(v) Both lamps light normally. If one is removed the other stays on.

(vi) Both lamps light brightly. If one is removed the other stays on.

(vii) Lamp (a) lights brightly: (b) and (c) light normally. If (a) is removed, (b) and (c) stay on. If either (b) or (c) is removed singly, the other lamp, (b) or (c), goes out but (a) remains on.

For the lamp used in this experiment, the correct voltage causes the correct electrical current to flow through the lamp and make it light normally. This happens in circuit (i) but in circuit (ii) the extra voltage causes more current and the lamp becomes brighter. It could 'burn out' as a result.

In circuit (iii) the lamps are connected in series. The voltage is 'shared' by the lamps so each has only 0.75 V; this gives less than the right current required, so both lamps will be dim. Taking out one lamp 'breaks the circuit' and no current will flow. In (iv), where two cells are used, each lamp gets the correct voltage (1.5 V) and the correct current flows in each lamp. Again, removal of one lamp breaks the circuit.

In (v) and (vi) the lamps are connected in PARALLEL. Current does not have to pass through one lamp to reach the other so removal of one lamp does not cause the other to go out. In (vi) both lamps are connected to 3.0 V so both are bright.

Circuit (vii) shows two lamps (b) and (c) connected in parallel with lamp (a). Lamp (a) is connected directly to 3.0 V and is bright; lamps (b) and (c) share 3.0 V and each lights normally.

Series connection of lamps is rarely used except for Christmas tree lights. Twenty small 12V lamps connected in series to a 240V supply would each receive 12V and would light normally but, if one were to be removed or to burn out, all the rest would go out (see Figure 115).

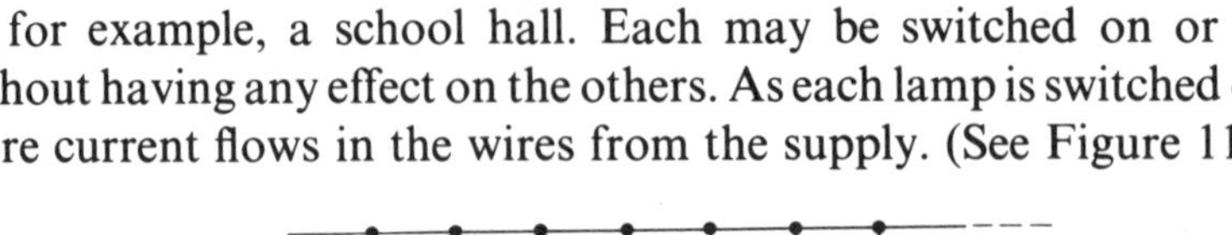

Figure 115

Parallel connection of lamps is used where there are many lamps in, for example, a school hall. Each may be switched on or off without having any effect on the others. As each lamp is switched on, more current flows in the wires from the supply. (See Figure 116.)

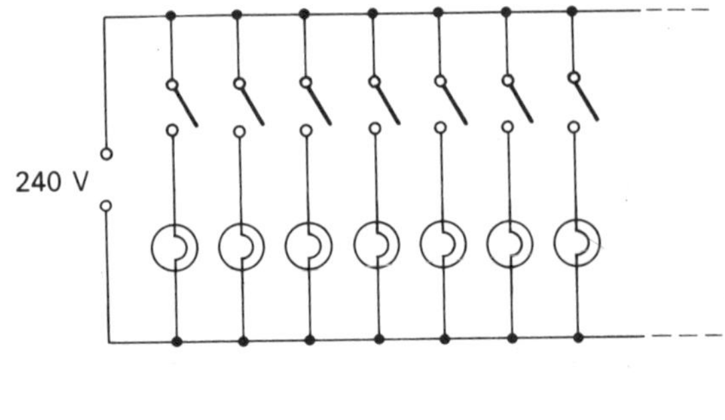

Figure 116

Electrical current, the 'flow of electrons' is measured in AMPERE (A) with a meter called an ammeter. In Figure 117, in each of the circuits, *all* of the current going through the lamps is measured by the ammeter. In Figure 117 (a) this would be 0.2 A, in (b) 0.1 A and in (c) 0.4 A. Each lamp filament, a very fine piece of wire, has a certain electrical resistance to the flow of current. In (b) where the lamps are in series, the current has to go first through one lamp and then the other so that the resistance is *increased.* The single cell has not enough pressure (voltage) to make the right current flow through that resistance to light each lamp normally and so the current is reduced. In (c), the fine wires are 'side by side' so the current has two paths to follow which will allow the right current to flow through the lamp. The *total* resistance is less and so more current flows.

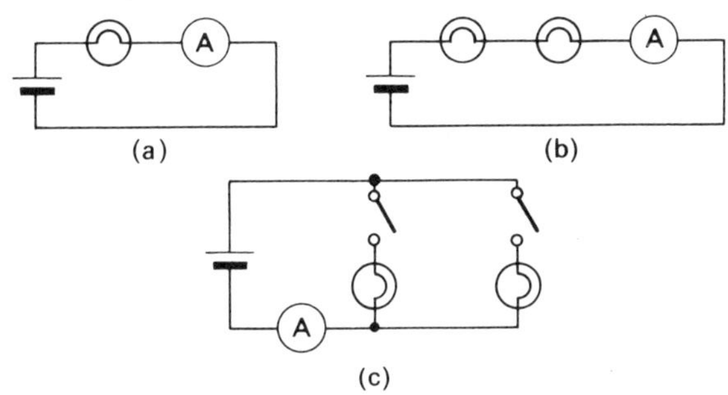

Figure 117

The current flowing through a lamp may be controlled to make the lamp bright or dim as wished. One way to do this is to place in series with the lamp a long length of fine wire which adds to the resistance. If the length of this wire is shortened, the resistance becomes less and more current will flow. This is usually done by winding the wire on a circular former which has a metal 'brush' which can be moved around making contact with any length of wire as required. (See Figure 118(a).)

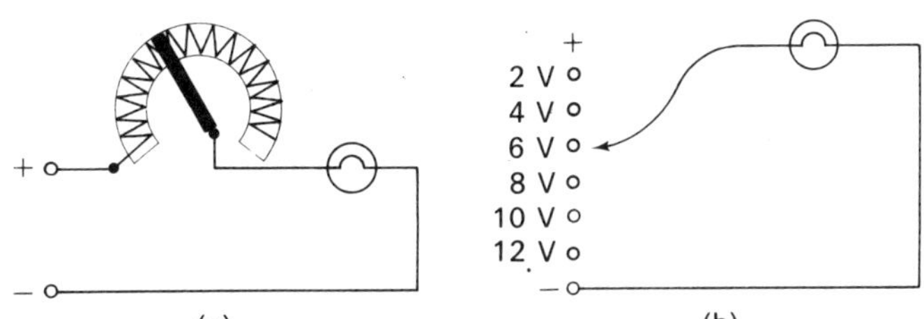

Figure 118

Another way of controlling current is to vary the voltage of the supply to the lamp. This can be done by 'tapping' different voltages from a battery or a special power unit (Figure 118(b)). The greater the voltage, the greater the current in the circuit.

Electrical power

'Power' in electricity means the same thing as it does with any form of energy: the 'rate of energy transfer' – how many joules of energy are transferred per second. In electrical circuits, the power is calculated by multiplying the voltage (V) by the current in ampere (A) giving the power in watt (W):

$$\mathrm{W} = \mathrm{A} \times \mathrm{V}$$

A mains electric lamp of 100W working on the mains of 240V takes a current of about 0.4 A (240 V × 0.4 A = 96 W).

A car headlamp working on 12V and taking 4A current would be 48W rated. A single bar electric fire and an electric kettle are often rated at 1000W (or 1 kilowatt (kW)) and take a current of about 4A.

Fuses

Any electrical current produces heat. A large current, if it is flowing in too thin a wire, can produce enough heat to melt the insulation of the wire and perhaps cause fire. This is prevented by placing in the electrical circuit a *fuse*, a piece of thin wire which, if a certain current flows through it, will melt and thus break the circuit. Such fuses are found in electric plugs (Figure 119).

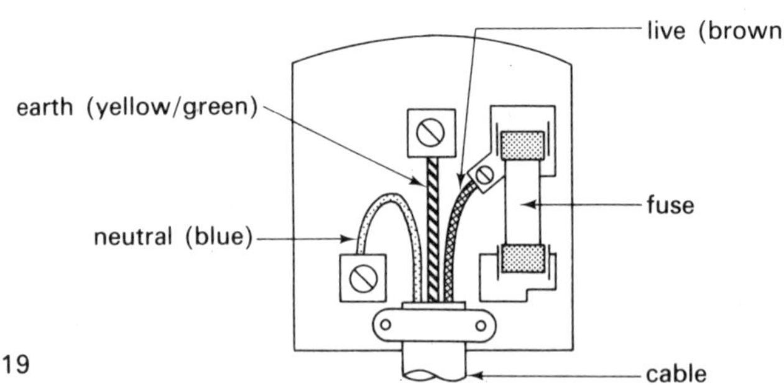

Figure 119

It is most important that the plug is correctly connected. The fuse must be in the LIVE line of the circuit and all switches must also be placed in the live wire. The EARTH wire is, in fact, connected to the earth itself at a power station. Should, by accident, the live wire touch the metal casing of some electrical appliance, a person touching that would get an electric shock. If this does happen in a properly earthed appliance, a big current immediately goes from the live wire because of the low resistance of the circuit and the fuse promptly melts, breaking the circuit.

Generating electricity

Batteries are useful for providing electrical energy for flashlamps, radios and so on, but they cannot supply the huge amount of electrical energy needed in homes and industry. The generation of electricity in power stations is due to discoveries made by Michael Faraday from his experiments.

Having found that an electrical current would produce a magnetic effect, Faraday wondered whether a magnet might produce electricity. The experiment was carried out by placing a magnet inside a coil of wire, connected to a meter which would show if a current were to flow. (See Figure 120.)

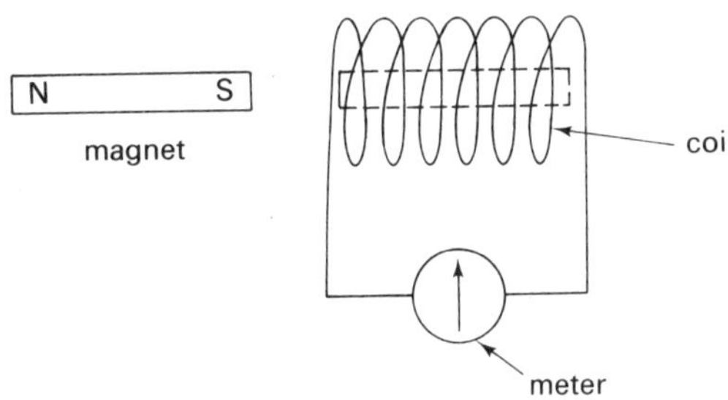

Figure 120

No current flowed when the magnet was at rest inside the coil but, as the magnet was pulled out, a current flowed for a short period of time in one direction and then, when the magnet was pushed into the coil, a current flowed for a short time in the opposite direction. Moving the magnet rapidly in and out of the coil produced an ALTERNATING current, one which flowed rapidly backwards and forwards in the circuit.

This effect is called ELECTROMAGNETIC INDUCTION. It is a way of turning kinetic energy into electrical energy.

A dynamo as used on a bicycle is shown in Figure 121. The turning of the magnet causes a magnetic field to change rapidly inside the coil producing an alternating voltage. The frequency of the voltage, the number of times it changes direction every second, and the voltage itself both increase as the magnet turns round faster.

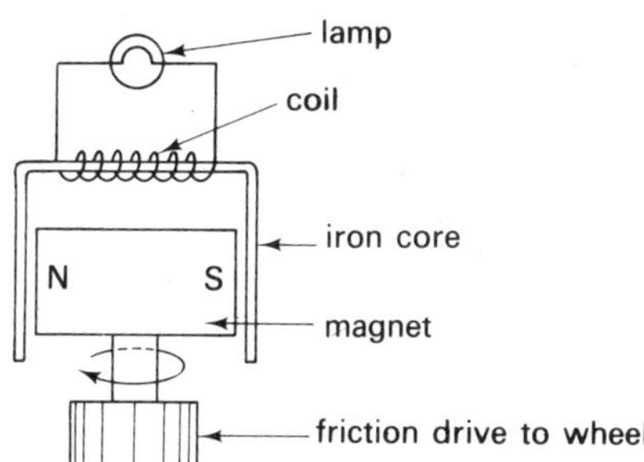

Figure 121

The generation of electricity in power stations changes heat energy, or kinetic energy from flowing water (hydro-electric power), into electrical energy by using huge dynamos or generators similar, in principle, to the simple cycle dynamo.

Self-test

CELL CONDUCTOR INSULATOR ELECTRON CURRENT VOLTAGE MAGNETIC HEATING CHEMICAL SERIES PARALLEL RESISTANCE AMPERE ELECTRICAL FUSE ELECTROMAGNETIC INDUCTION

Using the words in the list above, copy out and complete the sentences A to J:

A A material through which an electric current will pass is called a ______ .

B For safety, most electrical cables are covered with a material which is an ______ .

C A ______ changes chemical energy into ______ energy.

D Whenever an electric current flows in a wire it has a ______ and a ______ effect.

E An electrical current is a flow of ______ and is measured in ______ .

F An increase of current in a circuit is obtained by an increase in the ______ of the supply connected to the circuit.

G Electrical power is measured in watt and can be found by multiplying the ______ in ampere by the ______ .

H When connected to a battery, two lamps connected in parallel will take more current than the same two lamps connected in ______ because the ______ is less.

I A ______ is a thin piece of wire which will melt if too great a ______ flows in a circuit.

J Generating electricity by a moving magnetic field uses the principle of ______ discovered by Michael Faraday.

Questions

1 Draw circuit diagrams to show a battery of 6 V connected to (i) two lamps in series and (ii) two lamps in parallel. If the lamps were made for use on a 6 V supply, how would you expect them to appear in use in each case?

2 Supplied with 6 V lamps, four cells and an ammeter, describe experiments to show that (i) an increase in voltage causes an increase in current and (ii) an increase in resistance causes less current to flow if the voltage remains the same.

3 Describe simple experiments to show (i) the magnetic, (ii) the heating and (iii) the chemical effects of an electric current.

4 When using an electric iron explain the importance of (i) using

connecting wire of the correct thickness, (ii) connecting a fuse in the *live* wire and (iii) connecting an earth to the metal casing of the iron.

5 Describe a simple method by which it would be possible to find out whether various materials would or would not conduct an electrical current.

6 Draw a labelled diagram of a circuit to show how a number of 16 V lamps could be used to operate normally from a 240 V mains. Explain how, if a lamp failed, you could find out which one it was.

7 Explain how you could make a simple electromagnet using a six-inch nail and a length of insulated wire. How could you show the effect of reversing the direction of the current?

8 Describe carefully all the precautions to be taken to connect a three-pin mains plug correctly. *Without* using a mains appliance, how could you check that the fuse in the plug is in working order?

9 Draw a diagram to show a simple form of bicycle dynamo. Explain the *two* effects on the current produced if the dynamo is turned faster.

10 Two values of fuses are commonly used: 3 A and 13 A. Show, by calculation, why a 13 A fuse should be used on an electric kettle of 1000 W working on a 240 V supply.

12 Magnetism

It is often said that a magnet will attract 'metals'. This is not so. A magnet will attract iron, steel, nickel and cobalt, but it will not attract other metals such as copper, lead, aluminium or zinc. Magnetism can be found as a very weak force in many metals but it is strongest in metals like iron which are called ferrous metals. The metals which are attracted by a magnet can also be made into magnets.

If a magnet is hung up from its centre by a fine thread and left to swing freely, it will come to rest so that one of its ends – the north pole – points to north. This is not the Earth's true north but MAGNETIC north. It has been found that magnetic north has changed its position over hundreds of years. If another magnet is brought close to the hanging magnet, so that its north pole is brought towards the north pole of the hanging magnet, the hanging magnet is pushed away or REPELLED. If the south pole of the other magnet is brought close, the hanging magnet is attracted. This effect is summed up as:

LIKE POLES REPEL: UNLIKE POLES ATTRACT

The fact that a magnet will line itself up so that one end points to magnetic north, has been of great importance in navigation, how people find their way about the seas and lands of the world, and the land's magnetism is thought to have been known first to the Chinese. It was probably found in lodestone, a rock which has weak powers of attracting small pieces of iron.

The attraction of a steel nail or screw by a magnet can be explained by a theory that the magnet, when it gets close to the nail, makes the nail itself a magnet so that it has a south pole opposite the north pole of the magnet and, since unlike poles attract, the nail is drawn towards the magnet (see Figure 122).

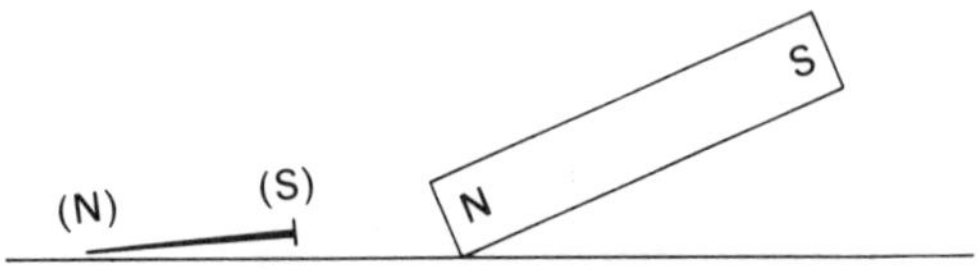

Figure 122

Since a magnet has an influence on iron objects and other magnets some distance away from it, it would seem that there is a FIELD OF FORCE surrounding it. Magnetic fields may be 'mapped' in several

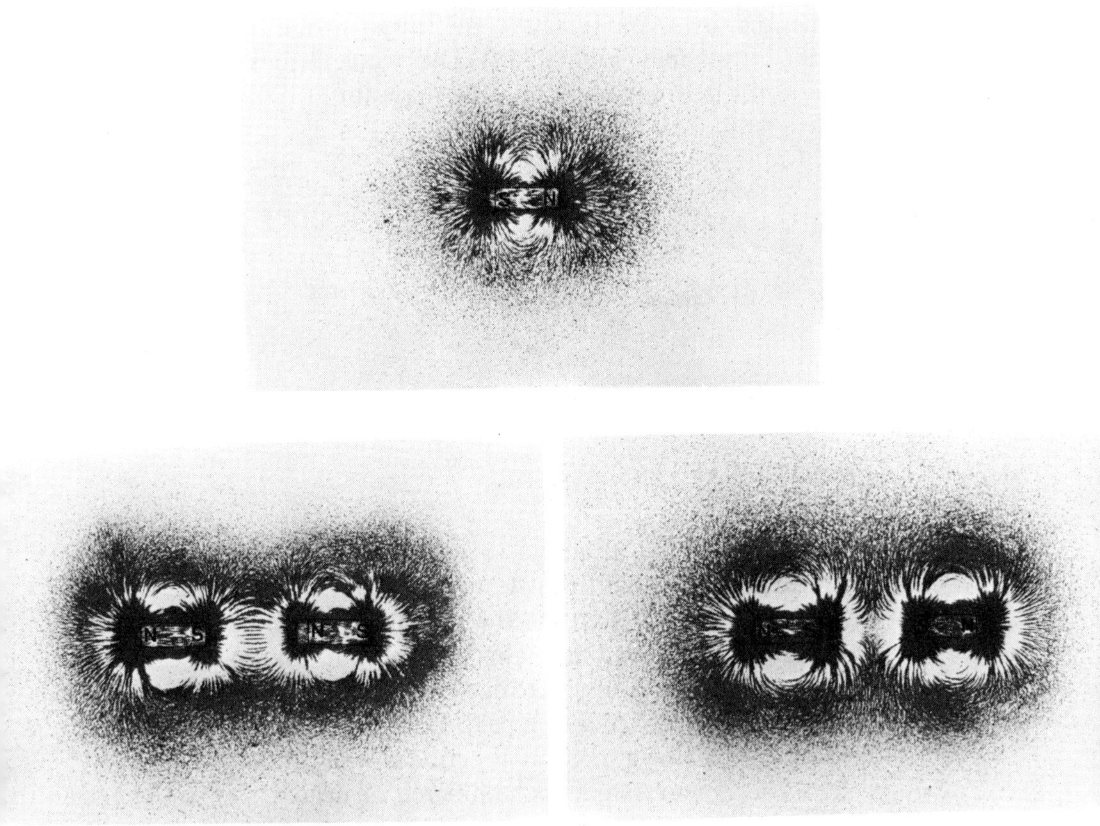

Figure 123

ways. A simple method is to place a magnet, or magnets, beneath a sheet of paper and then to sprinkle fine iron filings on to the paper. The iron filings begin to gather along curving lines. Gentle tapping on the paper with a pencil makes the patterns clearer. Examples of magnetic fields plotted in this way are shown in Figure 123.

It is important to realise that the LINES OF FORCE do not occur only where they appear on the paper but are found all around the magnet. When drawing diagrams of lines of magnetic force, their DIRECTION is marked in the way a compass needle would line up, with its *north* pole pointing towards the *south* of the magnet.

A bar of pure (soft) iron can be made magnetic by an electric current (see chapter 11). The soft iron loses its magnetism when the current in the coil is switched off. It does not RETAIN its magnetism. Steel, which is an alloy of iron and other metals and, in particular, special materials like cobalt steel, ticonal and magnadur, do retain the magnetism and can be made very powerful permanent magnets.

An unmagnetised piece of steel such as a needle may be

magnetised by stroking it several times in one direction with a permanent magnet (Figure 124). The needle is magnetised with its poles in the position shown in the diagram.

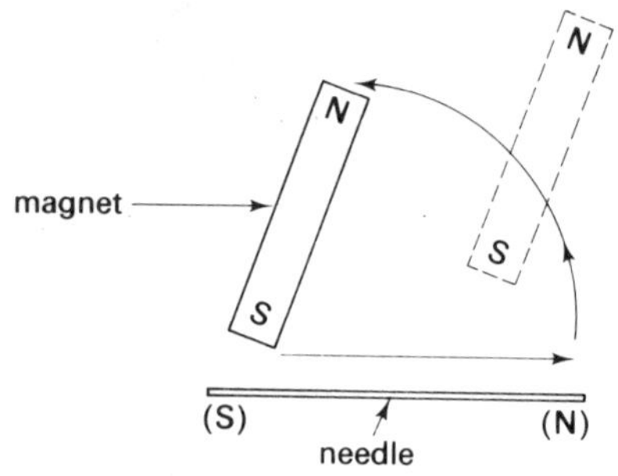

Figure 124

A theory which attempts to explain how a material may be magnetised states that the little 'atomic magnets' of metals like iron (the 'ferrous' metals) are, in the unmagnetised metal, arranged in small groupings called 'domains' as shown in Figure 125(a). These are shown at various angles to each other throughout the bar. When magnetised by a magnetising field, these tiny magnets straighten up and form into line along the direction of the field (Figure 125(b)).

If a permanent magnet is hammered or heated, the straight line pattern is broken up and the 'atomic magnets' tend to return to their domain pattern, producing a weakening of the magnet.

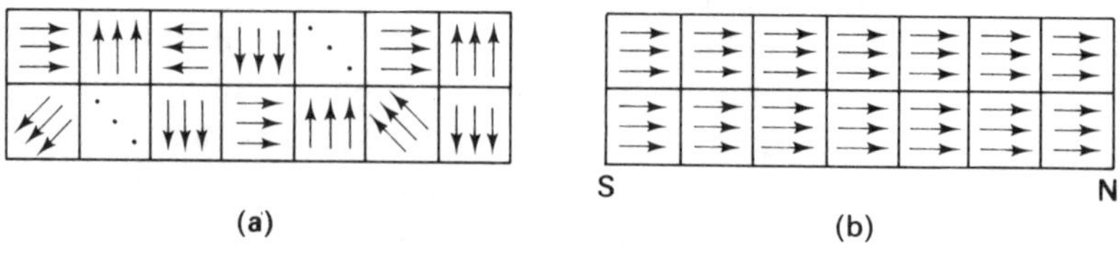

Figure 125

Testing for magnetism

A compass needle may be used to find out whether a bar of iron or steel is magnetised, but the test must be done carefully. It is not sufficient if the compass needle is ATTRACTED towards the bar. The bar might not be magnetised but the compass needle could move towards it all the same. To be certain that the bar is magnetised, it is necessary to see if one end of it will REPEL one end of the compass needle.

Self-test

FIELD COMPASS NORTH SOUTH LIKE
UNLIKE IRON STEEL NICKEL COBALT RETAIN
ALLOY ATTRACT REPEL HEATING MAGNETIC

Using the words in the list above, copy out and complete the sentences A to H:

A A compass needle points towards ______ north.
B ______ poles of magnets repel: ______ poles ______ .
C Permanent magnets may be made from steel but not from ______ because it does not ______ magnetism.
D The magnetic ______ surrounding a magnet may be traced with iron filings or by using a ______ .
E Apart from steel, ______ and ______ are other metals which can be magnetised.
F Ticonal is an ______ from which very powerful permanent magnets are made.
G Magnetism may be lost if a magnet is hammered and it is also weakened by ______ the magnet.
H The ______ pole of a compass needle will point towards the ______ pole of a magnet.

Questions

1 Describe a method of plotting the magnetic field of a bar magnet. Sketch the shape of the field formed when like poles of two magnets are close to each other.
2 Explain how a steel needle may be magnetised by using a permanent magnet. How could the needle be tested to find out which end of it is a north pole?
3 Explain, with diagrams to illustrate it, the theory which can account for an unmagnetised rod being magnetised and demagnetised.
4 Describe a simple method by which the *strength* of different, fairly weak, permanent magnets might be compared.
5 Explain how a compass is used to find out whether a steel needle has been magnetised.

13 Sound

Sounds are produced by things which vibrate, that is, move forwards and backwards at different speeds. Sound may be made by holding a ruler on a table so that some of it is beyond the edge of the table, raising the end of the ruler that overlaps and letting it spring back (Figure 126(a)). Whether the sound made is high or low depends upon the length that vibrates: the shorter this is, the higher the sound. The word PITCH is used to describe a sound: high pitch is a high note and low pitch a low note. Figure 126 shows other examples where vibrations produce sound. In (b) is a thin, stretched wire. Plucking the length AB between the 'bridges' causes a note of higher pitch than does plucking the length BC. Sound can be made by blowing across the top of test tubes in (c) causing the column of air inside the tube to vibrate. The shorter column of air in the tube (i), which has some water in it, produces the higher pitch. Tuning forks in (d) are made to vibrate by tapping the springy steel arms. The fork with the longer arms produces the lower pitch. The toothed wheel in (e) is turned around and the teeth make the thin piece of card vibrate. The faster the wheel is turned, the higher the pitch produced.

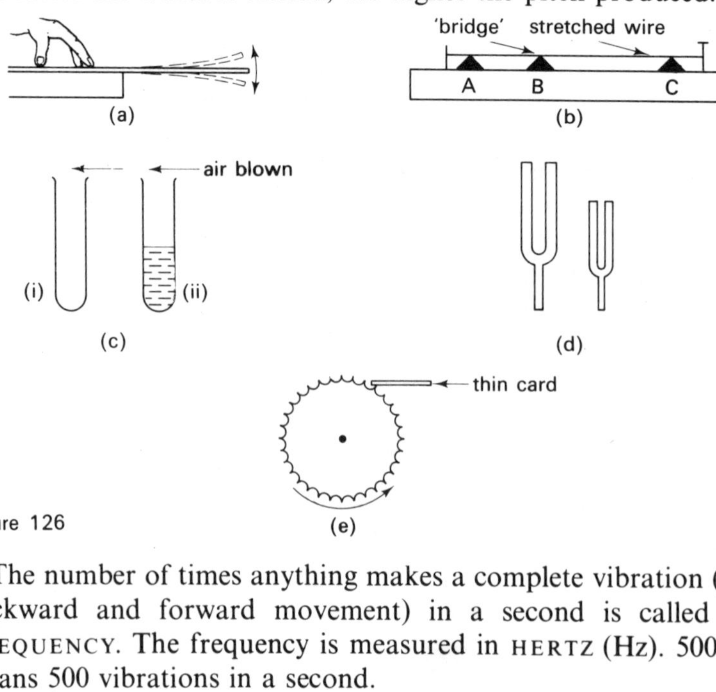

Figure 126

The number of times anything makes a complete vibration (one backward and forward movement) in a second is called the FREQUENCY. The frequency is measured in HERTZ (Hz). 500 Hz means 500 vibrations in a second.

In Figure 126(e), the card makes the highest pitch sound when it is made to vibrate quickest. PITCH depends upon FREQUENCY: the greater the frequency, the higher the pitch.

All the sounds made by the methods shown in Figure 126 can be heard in a room, so sound must travel through the air. Sound also travels through liquids and solids.

Figure 127 shows two simple experiments. In (a) a watch is sealed in a watertight tin and placed at the bottom of a beaker of water. If the ear is placed at the end of the glass tube dipping into the water, the ticking of the watch can be heard. People swimming under water can often hear the sound made by the propeller of a ship quite a long way away.

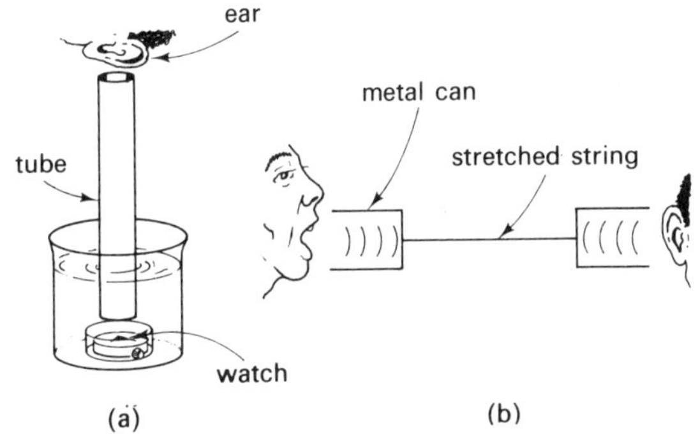

Figure 127

Figure 127(b) shows a toy 'telephone'. Small holes are made in the bottom of two metal cans and a long length of thin string is stretched between them. The sounds made by a voice come from the vibration of the vocal cords in the throat. These vibrate the air which in turn vibrates the metal bottom of the can. The vibrations then travel along the string, cause the other can to vibrate and these vibrations are carried through the air to the ear.

Since the vibrations travel through gases, liquids and solids, it could be that the MOLECULES making up these materials vibrate, the movement of one making the next one move in turn. This theory could be checked by finding out whether sound will travel through space where there are no molecules to vibrate. Figure 128 shows an experiment to do this. An electric bell is hung inside a bell jar by a rubber band. Wires are passed through the rubber bung and, when the current is switched on, the sound of the bell ringing can be heard. The air inside the bell jar is then removed with a vacuum pump. The sound of the bell becomes fainter and eventually stops even though the hammer can still be seen to be striking the bell. Sound cannot travel through a vacuum: it requires a MEDIUM to carry the vibrations from one place to another.

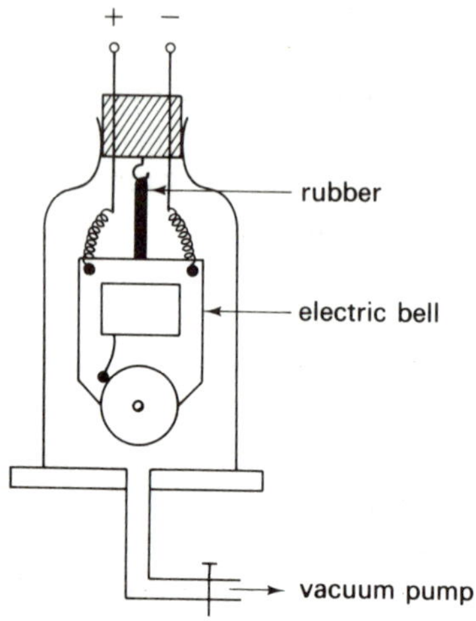

Figure 128

In much the same way that light is reflected from suitable surfaces like mirrors, sound vibrations are reflected and cause echoes. Light however can pass through space since, for instance, light reaches us from the sun across 150 million kilometres (93 million miles) of space. Light has been found to travel at the speed of 3×10^8 m/s (186000 miles per second) but sound does not travel at anything like this speed.

When a starting pistol is fired at some considerable distance from a person, the puff of smoke coming from the pistol is seen first and the sound of the explosion is heard a little later, even though the explosion and the smoke occur at the same time. During a storm, the flash of lightning is seen and the sound of the thunder may be heard seconds later depending upon how far away the storm is. The lightning and the thunder happen at the same time as the electrical charge on the thunder cloud causes the gigantic 'spark'.

An idea of the speed at which sound travels in air can be obtained from an experiment using a starting pistol and a stop-watch (Figure 129). A person standing a measured distance from someone firing a starting pistol, starts a stop-watch when the flash or puff of smoke is seen and stops it when the sound is heard. The stop-watch records the time in seconds which the sound has taken to travel the measured

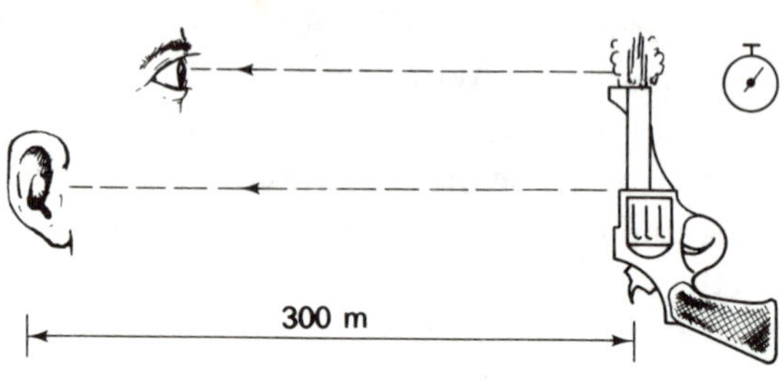

Figure 129

distance. The speed can be found by dividing the distance by the time. Sound travels in air at about 330 m/s (about 700 m.p.h.). Wind can affect the results of such an experiment and it is best done on a calm day.

Sound travels at different speeds in solids and liquids, the speeds being higher than through air. When anything, like an aircraft, travels at a speed greater than that of sound, it is travelling at a SUPERSONIC speed. This causes shock waves in the air making the 'booms' when an aircraft 'goes through the sound barrier'.

Transmission and recording of sound

The vibrations of air molecules caused by sounds can be made into electrical currents which vary as the sounds do in frequency and 'loudness'. These currents may be carried long distances along wires and then used to make something like a telephone earpiece vibrate which, in turn, produces vibrations in air which can reach the ear. Figure 130 shows a microphone of thc type used in telephones. It contains grains of carbon which the sound waves compress causing their resistance to change. The current from the battery changes because of this. At the receiving end, the varying currents passing through an electromagnet make a thin piece of iron vibrate.

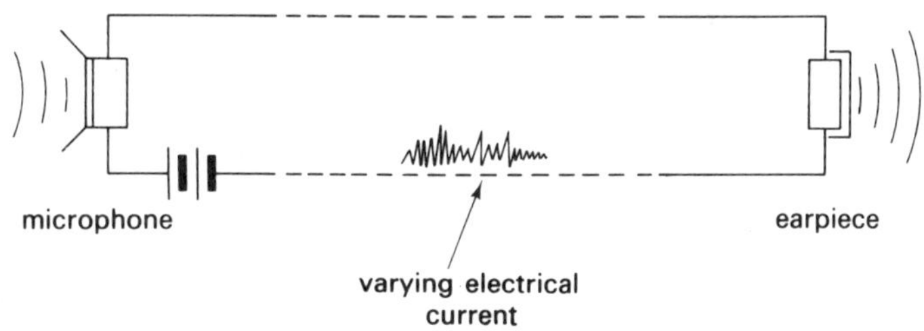

Figure 130

The gramophone disc has complicated patterns in its groove which cause the stylus of a pickup to vibrate. Inside the pickup, these vibrations are made to generate a very small varying electrical current. Passed to an amplifier, these very small currents are made much more powerful and the varying magnetic field they make causes the cone of a loudspeaker to vibrate. (See Figure 131.)

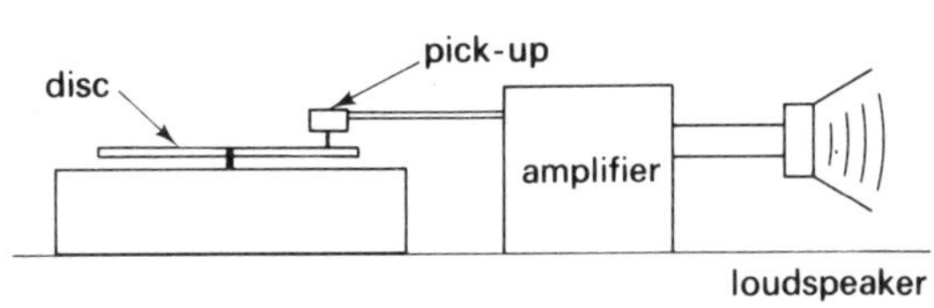

Figure 131

Self-test

HIGH LOW PITCH FREQUENCY ECHOES TENSION
RAISED LOWERED SPEED VIBRATIONS
AIR WATER LIGHT HERTZ VIBRATE

Using the words in the list above, copy out and complete the sentences A to H:

A All sounds are produced from ______ .
B The number of complete vibrations which a violin string makes in one second is called the ______ and is measured in ______ .
C A ______ pitch note has a higher ______ than a low pitch note.
D The pitch of the note made by a stretched string can be raised by increasing the ______ of the string.
E Sound waves are reflected and cause ______ .
F The speed of sound in ______ is greater than it is in air.
G A telephone works because varying electric currents in the wires make a thin iron plate ______ .
H If the length of a stretched, vibrating string is shortened, the pitch of the note it produces is ______ .

Questions

1 Describe an experiment which shows that sound cannot travel through a vacuum.
2 Describe any experiment which shows that the pitch of a note is raised if the frequency of vibration increases.
3 Describe an experiment which can be carried out to find the speed at which sound travels in air.
4 Explain how notes of different pitch are obtained in (i) a guitar, (ii) a trombone and (iii) a flute.
5 Explain briefly how sound is reproduced by means of a gramophone.

14 The kinetic theory of matter

A scientific theory is an idea formed to try to explain some of the things which scientists notice as a result of observation and through experiments. A theory has to be very thoroughly tested to see whether it holds good and it is necessary to carry out a very large number of additional experiments for this purpose. If a theory is right, it helps a scientist to PREDICT the result of an experiment, to be able to say what is most likely to happen before the experiment is carried out.

Here is a list of things which have already been learned from observations and experiments.

(i) There are three states of matter: solid, liquid and gas.
(ii) Changes of state can be made by changes in temperature.
(iii) Solids, liquids and gases all expand when they are heated.
(iv) When a gas is heated in a sealed container, its pressure rises.
(v) Heat travels slowly through a good heat conductor, from one end of a bar of metal to the other.
(vi) Changes of state require heat energy which does not change the temperature of the material. When a liquid changes into a vapour, it takes heat energy from its surroundings and causes cooling.

In order to explain why all these things happen, scientists developed a theory that all matter is made up of very tiny particles which are always moving. This is called the KINETIC THEORY OF MATTER and says the following.

(i) All matter is made up from very tiny particles.
(ii) These particles are always moving: very little in the case of solids – just vibration. They vibrate more in liquids and in gases they move about very rapidly.
(iii) In solids the particles attract each other strongly. In liquids, the attraction between them is weak and in gases they do not attract each other.
(iv) The movement of the particles increases when the temperature rises.

Figure 132 shows how these particles called MOLECULES are arranged in a solid, a liquid and a gas. (a) is a solid. Although vibrating, the molecules attract each other strongly. A solid can hardly be compressed at all and it keeps its shape. (b) is a liquid. The forces of attraction between the molecules are weak. A liquid can hardly be compressed at all. It does not hold its own 'shape'. (c) is a

Figure 132

(a) (b) (c)

gas. The molecules are widely spaced and move about very rapidly. A gas can be compressed – the molecules can be forced closer together. In a closed container, when the molecules are forced into a smaller space they will hit the walls of the container much more often. This 'bombardment' explains the rise in pressure.

The kinetic theory of matter could account for these facts. The kinetic theory can also be used to explain further experimental observations.

The motion of the molecules, their kinetic energy, could be increased by heat. The slow movement of heat along a bar of metal could be explained by increased vibration of the molecules spreading through them all. If the vibration increases, the space between the molecules would increase slightly and this would cause expansion.

When a solid becomes a liquid, some additional kinetic energy is needed to weaken the forces of attraction and, when a liquid changes into a vapour, a lot more energy is needed to give the gas molecules their great speed. This energy is the latent heat energy needed for changes of state.

The theory would be strengthened if it were possible actually to *see* the molecules and their movement but they are too small to be seen under the most powerful of microscopes. Certain experiments, however, show the *effects* of the movement of the molecules. Such an experiment is shown in Figure 133 where a strong solution of

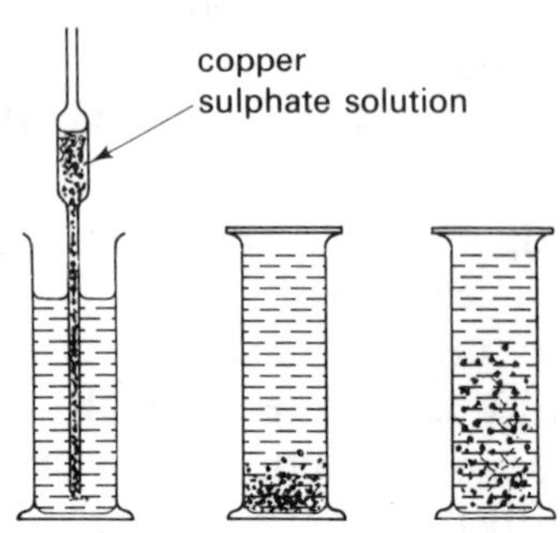

Figure 133

(a) (b)

copper (II) sulphate is made to form a dark blue layer at the bottom of a tall jar of water by letting it in gently with a pipette. The layer with clear water above is shown in A. The jar is covered and left undisturbed for several days. After that time the blue colour has spread through the water (B). This can be explained by saying that both water and copper sulphate molecules move quite slowly, and mix one with the other. This sort of movement is called DIFFUSION.

Another experiment is shown in Figure 134 where a little smoke from smouldering string is collected in a small glass cell which is covered with a thin glass disc. The cell is placed under a microscope and a beam of light is passed through the cell. When looked at through the microscope, the smoke particles are seen to move about in a rapid, jerky way. The smoke particles cannot move of their own accord. They are very big compared with molecules and are themselves made up of many millions of molecules.

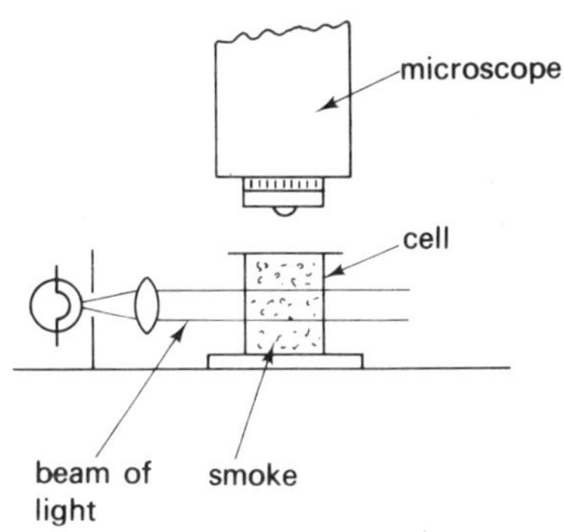

Figure 134

The movement is caused by the rapidly moving *air* molecules striking the smoke particles. This effect was first noticed when a scientist named Robert Brown, who studied plant life, was looking through a microscope at tiny grains of pollen in water. He noticed the pollen grains moving and could not, at first, understand why this happened. The pollen grains moved because they were being struck by the moving, but invisible, water molecules. The effect is now called Brownian motion.

If the kinetic energy which makes molecules move is obtained from heat, it would seem that molecules should move less as heat energy is removed. When a gas is cooled, it contracts as it loses kinetic energy and the molecules move less rapidly. It would seem that there should be a point when, without any kinetic energy at all, the molecules should stop moving.

Experiments have shown that all kinetic energy of molecules should be lost at a temperature of $-273°C$. Although such a temperature has not been reached, it is used as the starting point of a temperature scale called the Kelvin (or 'absolute') scale. The word absolute in this sense means a limit: it is not possible to go below 'absolute zero'.

The Kelvin scale of temperature can be calculated by adding 273 to the Celsius scale. 0°C becomes 273 K (° is not used in writing the Kelvin scale temperatures). Similarly 100°C is 373 K and $-20°C$ is 253 K.

Self-test

MOTION MOLECULES GAS SOLID LIQUID PRESSURE INCREASE ATTRACTION HEAT TEMPERATURE MATTER KINETIC ABSOLUTE KELVIN

Using the words in the list above, copy out and complete the sentences A to H:

A The ______ theory of matter says that matter is made up of ______ which are always moving.
B In solids there is a strong force of ______ between the molecules.
C The vibrations of ______ ______ when the temperature is raised.
D The increase of ______ of a gas heated in a closed container is due to the ______ in speed of the molecules.
E When molecules have no ______ energy the temperature will have reached ______ zero.
F A liquid does not keep its shape because there is little ______ between the molecules.
G By adding 273 to the Celsius temperature the ______ scale temperature is found.
H Latent heat provides liquid molecules with the ______ energy required by high speed gas molecules at the same temperature.

Questions

1 Describe any experiment which illustrates the motion of molecules in a gas or a liquid.
2 Explain how the kinetic theory of matter accounts for the expansion of solids when they are heated.
3 How does the kinetic theory of matter account for the fact that gases are easily compressed but solids or liquids are not?
4 Suggest how a *model* of Brownian motion might be made using small and large marbles.
5 What is meant by the terms 'absolute zero' and 'Kelvin scale'? Change the following temperatures into Kelvin:
(i) 0°C (ii) 200°C (iii) $-100°C$.

Index